Springer Climate

Series Editor

John Dodson, Menai, Australia

Springer Climate is an interdisciplinary book series dedicated to climate research. This includes climatology, climate change impacts, climate change management, climate change policy, regional climate, climate monitoring and modeling, palaeoclimatology etc. The series hosts high quality research monographs and edited volumes on Climate, and is crucial reading material for Researchers and students in the field, but also policy makers, and industries dealing with climatic issues. Springer Climate books are all peer-reviewed by specialists (see Editorial Advisory board). If you wish to submit a book project to this series, please contact your Publisher (elodie.tronche@springer.com).

More information about this series at http://www.springer.com/series/11741

An Thinh Nguyen • Luc Hens

Human Ecology of Climate Change Hazards in Vietnam

Risks for Nature and Humans in Lowland and Upland Areas

Springer

An Thinh Nguyen
Research Institute for Resources
and Climate Change
Hanoi University of Natural Resources
and Environment
Hanoi, Vietnam

Luc Hens
Vlaamse Instelling voor Technologisch
Onderzoek (VITO)
Brussels, Belgium

ISSN 2352-0698 ISSN 2352-0701 (electronic)
Springer Climate
ISBN 978-3-030-06937-7 ISBN 978-3-319-94917-8 (eBook)
https://doi.org/10.1007/978-3-319-94917-8

Softcover re-print of the Hardcover 1st edition 2019

Printed on acid-free paper

This Springer imprint is published by the registered company Springer International Publishing AG part of Springer Nature.
The registered company address is: Gewerbestrasse 11, 6330 Cham, Switzerland

Prolegomena

When I got the invitation from the Editors to write a preface for a monograph studying human ecology of climate change hazards in Vietnam, I immediately expressed my interest in reading the essay. I found out that the selection of the place (Vietnam) to study risks for nature and the humans in a holistic way was so didactic and wisely chosen that I thought it could be used as a pilot study with applications in other parts of the world. The reasons have to do with the complex topography and the general geographic location of Vietnam in relation to the most dangerous extreme weather phenomena whose frequency and intensity have been increasing in the past few decades. In addition, human ecology in such a complex topography and climate indeed offers a unique opportunity toward original research of the physical and socioeconomic impacts of climate change to be applied both to the coastal and mountainous areas of Vietnam.

Professors An Thinh Nguyen and Luc Hens as Editors in Chief are providing to the scientific community a unique collection of original research papers and offer clarifications on the relationships between local communities and various environments subjected to climate change hazards. It is obvious to the reader that the aim of this monograph is to provide a holistic outlook on the impacts to the societies in Vietnam as affected at least partially by climate change. Human ecological principles have been applied to local communities concerning their response to a hazardous environment. The book provides an inspiring mosaic in a wide range of applications of human ecology. The diversity of biophysical, socioeconomic, and policy responses are interpreted in an easily readable way, and the monograph successfully provides information on the economic effects at the most affected areas of Vietnam.

It is interesting to note here that the selection of two types of Vietnamese landscapes, i.e., the dominant mountainous and the much smaller coastal lowland, can be used as local examples that other countries with similar topography can use on their own adaptation strategies to climate change hazards. It will be also extremely useful to Vietnam policies to cope according to the complexity of their landscape and the response of the society to extreme weather and environmental effects.

This book offers to the reader an interesting navigation in the effects of the most violent weather tropical storms, floods, and on longer time scales sea level rise and coastal erosion, highlighting the impacts on humans and their development prospects. Heavy rains, floods, and landslides and the biophysical and socioeconomic impacts at selected sites are impressively presented. Finally, the book provides useful recommendations toward resilient landscapes and improved green cities in Vietnam.

To conclude, this monograph is original, presents new results, and provides a new look in its conclusions clarifying relationships between local communities subjected to climate and weather hazards. Although it contains lots of useful information for this particularly vulnerable part of the world, the conclusions have significant global value embracing international interest to the reader. I am confident that the reader of this impressive monograph will find it as useful and as a pleasant exercise which applies also to other parts of the world similar to the Vietnam social and natural vulnerability to spectacular phenomena and to societies at complex topography. The authors and the Chief Editors should be very proud for the outcome of such a complex and difficult task that they have so successfully accomplished.

Athens, 5 July 2018

Christos Zerefos
President of the Hellenic Foundation
for Research and Innovation

Foreword

Over the last 25 years, the issue of climate change has left the confines of the scientific research community, and is now recognized as a major, global societal problem. Gradually and progressively, potential impacts on future generations are becoming understood, including essential aspects of the way society will produce energy and organize its mobility. These realizations have moved the debate from scientific uncertainty and its associated doubts, to concrete policy and societal actions.

Vietnam ranks among the countries most affected by climate change. Although there is a lot of variation in the country, which spans 15° latitude, temperatures have overall increased, rain regimes have changed, monsoon season has extended, and tropical storms have increased both in number and intensity.

These changes increasingly affect both the natural and the human environment. Storms cause erosion of the coastline and beaches which are necessary assets for the inhabitants of the dunes and for tourism. An increasing number of Vietnam's islands have already disappeared in the sea or are threatened by erosion. People living near the sea or in the lowland of the Red River delta and the Mekong delta face the consequences of the storms and inundations, and find their houses and the public infrastructure damaged or destroyed.

As a part of the United Nations Framework Convention on Climate Changes and in the context of its "green economy" shift, the Vietnamese government established a comprehensive climate change policy which is now being implemented by the provinces and the communes. Coastal cities such as Da Nang and Hai Phong have successfully implemented green innovations. They established a renewable energy policy, paid particular attention to the green patrimony of the city, and managed their ports effectively.

I particularly welcome this book because of its interdisciplinary character: It looks beyond the biophysical aspects of climate change and puts particular emphasis on people. It deals with their perception, their expectations, and their realizations adapting to the effects of climate change. It discusses the resilience and the innovation on the subject. As an economist, I particularly appreciate the sections on the monetary aspects of storms and natural disasters addressed by the book.

The book has a series of original aspects, not only its human targeted inspiration but also the human ecological approach. Combining different methods in original research frameworks is a strength which might provide inspiration to scientists worldwide. On its contents, the reader will find here not only an analysis of the already often addressed natural disaster problems along the coast, but also the less studied mountains which are frequently affected by storms.

This book deserves a wide reading audience. Although it is primarily a contribution to the human ecological science backing climate change, the content and the analysis are important for all of us concerned about sustainable development and the effects of the increasing frequency and intensity of the associated natural disasters. Allow me to strongly recommend this book not only to scientists but also to a wide range of societal actors.

Assoc. Prof. Dr. Nguyen Van Thanh
Vice-Minister Security, Vietnam

Acronyms

AHP	Analytic hierarchy process
ARIMA	Autoregressive integrated moving average model
CBA	Cost-benefit analysis
CCFSC	Central Committee for Flood and Storm Control
CI	Consistency index
CR	Consistency ratio
CRA	Community risk assessment
DEM	Digital elevation model
DPSIR	Driver-pressure-state-impact-response
EBM	Ecosystem-based management
EEA	European Environment Agency
EIA	Environmental impact assessment
EZ	Economic zone
FAO	Food and Agriculture Organization
FHZI	Flooding hazard zone index
GDP	Gross domestic product
GIS	Geographic information system
ICZM	Integrated coastal zone management
IMGG	Institute of Marine Geology and Geophysics (Vietnam)
IMHEN	Institute of Meteorology, Hydrology and Environment
IPCC	Intergovernmental Panel on Climate Change
ISPONRE	Institute of Strategy and Policy on Natural Resources and Environment
LHZI	Landslide hazard zone index
LULCC	Land use land cover change
MCDA	Multiple-criteria decision analysis
MONRE	Ministry of Natural Resources and Environment (Vietnam)
NPV	Net present value
NTP-CC	The National Target Program to Respond to Climate Change
OECD	The Organization for Economic Co-operation and Development
OPEC	Organization of Petroleum Exporting Countries
PTSD	Post-traumatic stress disorder

RCPs	Representative concentration pathways
RI	Random consistency index
SP-RCC	Support Program to Respond to Climate Change
UNDP	United Nations Development Programme
UNFCCC	United Nations Framework Convention on Climate Change
WB	World Bank
wMean	Weighted mean

Contents

Contributors

The Anh Luu VAST Institute of Geography, Hanoi, Vietnam

Quoc Anh Trinh VNU University of Sciences, Hanoi, Vietnam

Thanh Cong Nguyen SP-RCC Coordination Unit, Hanoi, Vietnam

Ha T. T. Le Hanoi University of Mining and Geology, Hanoi, Vietnam

Thanh Hai Nguyen University of Missouri – Columbia, Columbia, MO, USA

Luc Hens Vlaamse Instelling voor Technologisch Onderzoek (VITO), Brussels, Belgium

Huong T. T. Chu MONRE Department of Climate Change, Hanoi, Vietnam

LY Thi Pham VAST Institute of Geography, Hanoi, Vietnam

Hoang Nam Nguyen ISPONRE, Hanoi, Vietnam

An Thinh Nguyen Research Institute for Resources and Climate Change, Hanoi University of Natural Resources and Environment, Hanoi, Vietnam

Thi T. M. Tong Hanoi University of Natural Resources and Environment, Hanoi, Vietnam

Thu T. M. Pham Thuy Loi University, Hanoi, Vietnam

Thuy L. T. Hoang VAST Institute of Geography, Hanoi, Vietnam

Van Hanh Ta VAST Institute of Material Science, Hanoi, Vietnam

Van Hong Nguyen VAST Institute of Geography, Hanoi, Vietnam

Van Trung Nguyen Hanoi University of Mining and Geology, Hanoi, Vietnam

Short Bibliographical Note

Editors in Chief

An Thinh Nguyen acquired PhD with his focus on mountainous landscape ecology and geography at Vietnam National University, Hanoi, in 2007. He did his postdoctoral research on remote sensing and GIS at Kookmin University (Korea) during 2009–2010. In 2014, he was appointed as associate professor of geography and acted as the director of Research Institute for Resources and Climate Change (IRC), Hanoi University of Natural Resources and Environment (HUNRE), Vietnam. His research interests include geography of Vietnam, mountainous landscape ecology, and quantitative geography. An Thinh Nguyen is the Vice President of the International Association for Landscape Ecology in Vietnam (VN-IALE) and a member of the Vietnamese Association of Geographers (VAG).

Luc Hens graduated as a biologist and received his Ph.D. in Biology from the Vrije Universiteit Brussel (VUB), Belgium. Until 2010 he was a professor and chair of the human ecology department. He also lectures at the Technical University in Sofia (Bulgaria), the National and Kapodistrian University of Athens (Greece), the Sumy State University (Ukraine), and the Lisbon University (Portugal). He was a senior scientific adviser at the "Vlaamse Instelling voor Technologisch Onderzoek" (VITO), which is Belgium's biggest environmental research organization. He is currently retired as an emeritus professor. Professor Hens' specific area of research concerns the elucidation of interdisciplinary instruments for sustainable development. In this framework, he acted as the promoter of over 50 research projects. Luc Hens acts as an expert in environmental policy on several councils in Belgium. He is the European editor for the *International Journal of Environment, Development and Sustainability*.

Introduction

General Theme and Subject

Worldwide, few countries exist where the effects associated with climate change are as pronounced as in Vietnam. Until now, most of the international attention on climate change hazards has been focused on the Vietnamese lowlands and the coast – the economically most productive part of the country. This is only fair, as the country experiences increasingly frequent and intense tropical storms as well as cyclones, floods, sea level rise, coastal erosions, droughts, and salinization. This affects local livelihoods, land use, and migration along the coast. Agriculture is pushed back toward the hinterland. The life of people is affected in such a dramatic way that they have to relocate. Forecasts of the effects all point to further worsening during the years to come. If by the end of this century the sea level rises by 1 meter, 10% of the population of Vietnam's coastal area would be directly affected, which reflects 80% of the country's GNP (MONRE 2016).

However, mountains cover over 75% of the total land of Vietnam (Le et al. 2012). They are also affected by climate change-associated events such as extreme weather conditions including tropical storms and heavy rainfall, which cause floods and landslides. This impairs small-scale agriculture, industry, tourism, infrastructure, livelihoods, and food security in an area shared by the Kinh with 54 other ethnic populations such as Tay, Thai, and Hmong minorities. Until now, much less attention has been paid to climate change-associated effects in these mountainous environments.

Understanding these complex climate change hazards, their impacts and adaptation, needs interdisciplinary frameworks. This book uses an inter- to transdisciplinary human ecological approach providing insight in the interrelations between changing land use, livelihoods, and migration in heavily affected areas along the Central Coast and in the northern mountains of the country.

A Human Ecological Approach

Complex problems of landscape change and environmental and social impacts and responses cannot be understood from a single, disciplinary perspective. This book adopts a human ecological approach which aims at understanding human-environment interactions in an interdisciplinary way. This allows integrating physical environmental effects with their impacts on livelihoods, culture, health, and perception, to mention just a few aspects. To this end, human ecology uses methods which combine and integrate scientific data. These methods stem from geosciences (GIS, remote sensing, spatial models), human sciences (Delphi survey, perception studies, interviews, local knowledge analysis, indicators), and applied sciences (environmental health and technology). A range of methods is applied in the studies documented in this book. This allows coming up with outcomes which go beyond disciplinary thinking about climate change and its associated effects.

Content

Research Areas

This book makes these general considerations in both regions of the country tangible. It offers a collection of original research papers on Ky Anh, the most southern district of the Ha Tinh province (along the coast of Central Vietnam, bordering the South China Sea), and on Van Chan, a mountain district of the Yen Bai province (in Vietnam's northern mountains). Both areas represent major landscapes of Vietnam: the lowland coast and the mountains. These latter cover over three-quarters of the country, while the lowland is the most important economic area.

The coastline of Ky Anh is perpendicular to the prevailing direction of the rising number of tropical storms, increasingly affecting the local inhabitants and counteracting their initiatives to build infrastructure and tourism development.

Van Chan is one of the centers of slope agriculture. This not only results in terraced rice fields, but this fragile environment is under increasing pressure toward higher yields and modernization. The region is equally in search of a sustainable model to handle increasing urbanization pressure. The protected forests and crop systems on the slopes of Van Chan are particularly important both locally and regionally. Van Chan and other mountains in the Vietnamese upland are the "sources" which directly affect the economy and the environment in the "sinks" of the Red River Delta. Therefore, in a global context of climate change, sustainable development in the mountains and the development of the lowlands are integrated and connected.

The chapters of this book clarify the relationships between local communities and environments subject to climate change hazards (or the coevolution between local communities and their changing environment). The results of the original studies are described and discussed in a scientific context. Research on climate change effects is hampered because it is often difficult to make a distinction between general environmental and specific climate change-related aspects and between the local and the more widely applicable aspects. Knowledge in this area proceeds by comparing effects in different ecosystems under a range of circumstances. Contributing to this discussion is one of the main merits of this book. In this way, this book allows international comparisons with similarly affected areas under continuing threats, which are or will be comparable with the situation in Ky Anh and Van Chan today.

Originality and Conclusions

This book shows unique elements that are complementary to the existing literature:

- The analysis of climate change-associated hazards and their effects on local communities according to human ecological principles. The latter provide a wider scope of integrated interpretation of biophysical, socioeconomic, and policy responses to climate change hazards in Vietnam.
- The comparison of climate change-associated effects along the Vietnamese coast and the less studied mountains of northern Vietnam, while both are mutually dependent.
- Monetary valuing of the climate change-associated effects in some of the most affected areas.
- The integration of local knowledge in adaptation (and mitigation).

Conclusions of this book relate to different geographical scales:

- Local: Data are collected in two specific types of Vietnamese landscapes. Therefore, results on climate change hazards and adaptations are of significance for the local mitigation and adaptation strategies and policies in Ky Anh and Van Chan.
- Regional and national: The two research areas have been selected because they stand for and to some extent represent the upland and the lowland in Vietnam. Both are interconnected as "source-sink" landscapes. The comparison of the results provides information on the differentiated measures and policies Vietnam needs to cope with in its various landscapes.
- International: The results of the local investigations are discussed in a general scientific context. This discussion among others contributes to distinguish between typical climate change effects and responses and between local and more general responses.

Structure

This book is structured in four parts. The first part introduces the background. The human ecological framework of this book is detailed. The geographical area and its topography are discussed illustrating the environmental and socioeconomic context of Vietnam's coast and mountains. In Chapter 1, methods are discussed illustrating the methodology aspects of human ecology; at the same time, these methods are used in subsequent chapters. Part two addresses the coast of Central Vietnam. The effects of tropical storms, floods, sea level rise, and coastal erosion in Ky Anh are described, highlighting the impacts on the local population and its development perspectives. Part three concentrates on the uplands of northern Vietnam, covering the effects of cyclones, heavy rains, floods, and landslides in the Van Chan mountain. The biophysical impacts are related with the socioeconomic effects. Part four makes policy recommendations in building resilient landscapes and green cities and discusses the potential implications of findings for practice in Vietnam.

The eight chapters of this book not only describe the effects of more frequent and intense climate change hazards, they also provide the wider context of integrated interpretation, socioeconomic implications, and policy responses. This necessitates original research strategies combining methods from the natural sciences with approaches which dovetail in the human sciences. This transdisciplinary human ecological context provides added value to this book: It offers methodological inspiration which can be used as a template to analyze similar situations worldwide.

Format of the Monograph

This is a complete book. All chapters include a preface, figures, tables, color maps, and photographs. Also notes on the contributors are provided. This book is concluded by an integrated list of up-to-date references. All the chapters of the core part are the result of original research, combining fieldwork, questionnaire results, GIS, spatial analysis, and statistics. Some 30% of the information has been published in two papers of the Springer *Journal of Environment, Development and Sustainability (ENVI)*:

1. Le, T., Nguyen Q.T., Nguyen M.T., Dang D.L., Nguyen V.P., Phi C.V., Bui X.D., Nguyen D.V., Vu V.C., Thanh N.L. (2012). Vietnam: land and people. Vietnam Education Publishing House, Hanoi, Vietnam. 544 pages.
2. MONRE (Vietnamese Ministry of Natural Resources and Environment) (2016). Climate change and sea level rise scenarios for Vietnam. Vietnam. 170 pages.
3. Nguyen, A. T., Vu, A. D., Vu, V. P., Nguyen, N. T., Pham, M. T., Nguyen, T. T. H., Le, T. H., Nguyen, V. T., Hoang, K. L., Vu, D. T., Nguyen, Luong, T. T., Trinh, P. N., Hens, L. (2017). Human ecological S.T. effects of tropical storms in

the coastal area of Ky Anh (Ha Tinh, Vietnam). *Environment, Development and Sustainability, 19*(2): 745–767.
4. Nguyen, A. T., Vu, A. D., Dang, G. T. H., Hoang, A. H., Hens, L. (2018). How do local communities adapt to climate changes along heavily damaged coasts? A Stakeholder Delphi study in Ky Anh (Central Vietnam). *Environment, Development and Sustainability, 20*(2): 749–767.

This book offers an opportunity overviewing the totality of the research and combining the results in a more interdisciplinary and complete scientific framework. The introductory and concluding chapters are written by the lead researchers of the projects on which the results are based.

Authors

The lead author of this book is Associate Professor Dr. An Thinh Nguyen who guided and performed an impressive series of geography, landscape ecology, and climate change hazard valuation projects in the study areas and is one of Vietnam's authorities. Currently, Professor Nguyen heads the Research Institute for Resources and Climate Change (IRC) at the Hanoi University of Natural Resources and Environment (HUNRE), Vietnam.

Professor Em. Dr. Luc Hens is a Belgian human ecologist who has been performing and supporting research on climate change and environmental policy instruments in Vietnam for over 25 years. He worked for the *Vlaamse Instelling voor Technologisch Onderzoek* (VITO), which is Belgian's largest environmental research organization. Professor Hens has an impressive list of over 150 publications in international, peer-reviewed journals. He currently acts as the editor in chief of the *Journal of Environment, Development and Sustainability (ENVI).*

Acknowledgment

This book is a partial fulfillment of the project NAFOSTED 105.07-2015.04. The authors are indebted to the households and the district and commune authorities who completed the questionnaires and provided opportunities for discussion on the results.

An Thinh Nguyen and Luc Hens
(Editors in Chief)

Part I
Background

Chapter 1
Human Ecology of Climate Change Hazards: Concepts, Literature Review, and Methodology

Abstract This introductory chapter provides the context and the scope of this book. It is structured in two main sections.

At first the motivation adapting an inter- to transdisciplinary, human ecological approach for the complex natural disaster problems is addressed. Among others, the consequences of this option are both a wide range of methods and the unavoidable uncertainty. The section on the human ecology of climate change hazards provides the conceptual model of the book: it combines the biophysical aspects with those of the human environment. A selection of the main effects of climate changes in Vietnam provides information on temperature trends that significantly varies over this long country. Next to temperature, also the variations in rain patterns and natural climates are introduced. Special attention is given to the coastal and mountainous study areas of this book.

Part two describes the methods used in the subsequent seven chapters. Of notice, a particular method may be used in different chapters. On the one hand, biophysical methods as remote sensing and GIS are described, with special reference to coastal erosion studies. On the other hand, methods which dovetail in the social sciences are discussed. They include the analysis of networks between stakeholders and monetary valuation studies. The human ecological way of thinking provides however particular attention to methods allowing the integration between biophysical and socioeconomic data. At this level, questionnaires, (variants of) Delphi studies, a combination of indicators and indexes, multi-criteria analysis, and spatial modeling are used.

The chapter is concluded by two tables indicating which methods are used in the seven remaining chapters of the book.

Keywords Human ecology · Climate change hazard · Time and space integration · Analytical complete · Combination methods · Vietnam

A. T. Nguyen, L. Hens, *Human Ecology of Climate Change Hazards in Vietnam*, Springer Climate, https://doi.org/10.1007/978-3-319-94917-8_1

1.1 Concepts and Literature Review

1.1.1 Human Ecology

Human ecology studies the relationships between people and their environment. Although this scientific approach dovetails in biology-related ecology, the idea emerged from the finding that almost all problems affecting the human environment find their origin in both biophysical and socioeconomic aspects (Steiner 2014). Or as pointed out by Nazareth (1996), human ecology is based on the hypothesis of the coexistence of "two systems in constant interaction: the human system (which receives and decodes information) and the environment that produces response actions." Human ecology evolves on the analysis and deep understanding of mechanisms of biology, geography, environmental and natural sciences, sociology, economics, and other human sciences, next to applied sciences as medicine and engineering. Human ecology not only integrates in its approach the (environmental, social, and economic) dimensions of sustainable development but equally pays attention to the cultural, psychological, and welfare dimensions of the equation. Today no scientific discipline from archeology and history, over law and policy, to elementary particulates and mathematics excludes its contribution to the understanding of the mutual human-environment interrelationship and interactions (Lawrence 2003; Pires et al. 2010). Human ecology is about the reciprocal relationships between the biophysical and the socioeconomic environment in its widest context. The in-depth analysis of the effects of extreme weather conditions on humans and their environment is the general aim of this book.

For the human ecologist, the environment is perceived as an ecosystem. The ecosystem is a wide concept both when it comes to its nature and its size. In view of its roots in biology, an ecosystem is everything in a specified physical area – air, water, and soil – which interacts with the biological organisms and physical structures: it is about the organisms in a lake which not only depend on each other but are also influenced by the water they swim in, the quality of the water bottom, and the air above the water surface. Evidently a forest, a mountain, and a coral reef are all (biological) ecosystems. Human ecology extends the ecosystem concept to areas in which humans are an important factor: a city, a farm, a tourism resort, and the cultural assets (Fig. 1.1).

Not only the broad nature of ecosystems characterizes the human ecological approach; also the time and space dimensions vary significantly. Geographical scales vary from household issues, over regional, national, and international, to continental and global scales. The human ecologist is interested in the integration of these wide varying levels. This book illustrates how global problems as climate changes deeply affect the regions, villages, communities, and households where the storms and droughts cause disasters.

Next to the scale, time is an important dimension for the human ecologist. From an environmental point of view, problems vary between a short frame of time (e.g., noise disturbance), over days floods last and growth seasons in agriculture, to

Fig. 1.1 Human ecology is about the interactions between the living, biotic environment (humans, ecosystems) and abiotic, resources environment (semi-intensive ponds for aquaculture in the Cat Ba Biosphere Reserve, an archipelago in Northern Vietnam). (Source: An Thinh Nguyen, 2008)

periods covering ten thousands of years (e.g., storage of high radioactive waste). From a sustainability point of view, the trans-generational time dimension matters. Also the effects of the climate change-associated extreme weather conditions vary in time between immediate (e.g., the flash flood), over years (e.g., erosion), to decades (e.g., migration and resettlement).

Part of the added value of human ecology is that it does not look into problems as a series of individual issues; rather it considers a web of interrelated problems. As illustrated in Fig. 1.1, climate change-related aspects provide an example in this respect: the biophysical aspects of greenhouse gas emissions are interrelated with the consequences as storms, drought, and sea level rise. On their turn they affect agriculture, business, industry, and tourism. These effects change the lives of people, who, guided by their perception, will adapt using a range of possibilities supported by policy. This book deals with this web of interdependent and interrelated problems.

Another part of the added value relates to the incommensurability and sometimes the incompatibility of dealing with problems in the environmental science realm. Different disciplines often result in different information and points of view. While the engineer will tend to advocate the (short-term) interests of industry and/or agriculture, the biologist will often come up for environmental protection and the long-term value of environmental services and assets. Human ecology provides frameworks bringing these different elements in line while respecting their heterogeneous character. This book provides a series of examples of how Delphi surveys succeed in reaching consensus on sensitive environmental, social, and economic issues among a variety of stakeholders.

This analysis of human ecology allows characterizing some of its essentials which make it particularly interesting for analyzing the natural hazards and extreme weather events associated with climate changes along the Vietnamese coast and in the mountains:

- *Multi- and interdisciplinarity:* Human ecology links the biophysical with the socioeconomic aspects of the human ecosystem. The approach provides integrated frameworks that allow more comprehensive, holistic understanding of the issues at stake. The multidisciplinary reach of human ecology is wide: it ranges from anthropology, over psychology and economics, to chemistry and biology, just listing these examples. This book offers a multidisciplinary analysis of the effects of storms on agriculture, on tourism, on natural ecosystems, on livelihoods, and on the quality of life of people. In doing so, an integral vision on the biophysical impacts, their effects on the (semi-)natural and sociocultural ecosystem, and the policy implications emerges. Its call for more balanced and responsible attitude toward the impacts of climate changes in a fast-developing country as Vietnam illustrates the philosophical-ethical dimension of the interdependence between the biosphere and the social sphere.
- *Wide scope of methods:* The multidisciplinary nature of human ecology results in using a wide range of methods. In general human ecological approaches use a mix of methods stemming from natural sciences, human sciences, and applied sciences. In view of the numerous, sometimes heterogeneous, data, integration methods as remote sensing and GIS and integration frameworks as the driver-pressure-state-impact-response (DPSIR) model are of core importance. Indicators and indices are most instrumental in this respect. Consensus among target groups with a wide variety of opinions on sensitive issues is reached using Delphi surveys, sometimes in combination with a DPSIR model. Results of questionnaires and standardized interviews are treated using conventional statistics for the social sciences. This book illustrates this wide range of methods used during field visits, data processing, and spatial modeling. Sea level rise and the resulting erosion of the coast are measured with the classical spatial analysis methods. On the occurrence of storms, meteorological methods provide information. On the quality of life, the livability under extreme weather conditions and the perception of the impacts, interviews, questionnaires, and checklists are valuable research instruments.
- Models help in integrating different layers of information. In this book the DPSIR model proved most useful in linking the biophysical basic data with effects and policy measures.

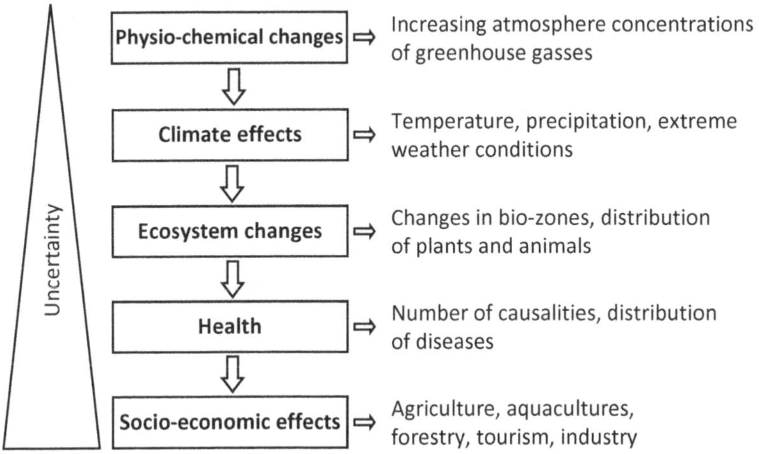

Fig. 1.2 The climate change complexity chain related to the degree of uncertainty

- *Uncertainty:* The different aspects of the wide scope of human ecology contribute in a variety of ways to uncertainty. The larger the geographical and the time scale of the problem(s) at stake, the higher the uncertainty. Monitoring the surface of a flood or the erosion of a beach can be done much more precisely than predicting the effects of climate change. Moreover these latter are also influenced by the nature of their interlinkages. As illustrated in Fig. 1.2, climate change provides an example. While the physiochemical changes as the increase of the greenhouse gasses in the atmosphere are described with the normal scientific rigor, it is already more difficult linking these basic drivers with changes in temperature, precipitation, storms, or extreme weather conditions. These result in ecosystem changes, which on their turn affect human and animal health and disturb the whole socioeconomic system. However in particular forecasting the quantitative effects at these different levels is also a matter of increasing uncertainty. This uncertainty should be acknowledged in the human ecological analysis: the larger the geographical scale and the time frame and the more complex the context of the problem, the higher the uncertainty. The more the uncertainty increases, the more the results are characterized by a postnormal science approach. This book recognizes the uncertainty linked with the results in the different chapters.
- The human ecological approach helps in understanding nowadays changes and future adaptations to climate change-associated disasters. It offers an integrated physical environmental, economic developmental, and managerial-legal perspective. This approach results in recommending balanced policy options for the future. The addressed complexity is not unique for Vietnam. The issues discussed in this book can be used as a template for a range of countries worldwide.

1.1.2 Human Ecology of Climate Change Hazards

Figure 1.3 has shown the conceptual model for human ecological approach to climate change hazard on which this book is based. The human ecological effects of natural hazards and extreme events are addressed as the impacts of natural hazards (in atmosphere, hydrosphere, geosphere, and biosphere) on human populations (noosphere). Hazards are natural processes which affect populations adversely (Peterson and Johnson 1995; UNDP 2007; Wang et al. 2012; Stoffel et al. 2016). Risk represents the relationship between humans and natural processes of environment. The issues in Vietnam are tropical storms, heavy rains, cyclones, extreme cold and droughts (atmospheric hazards), floods and flash floods (hydrospheric hazards), landslides (geologic hazards), insect infestations, disease epidemics, and wildfires (biospheric hazards).

A socio-ecosystem reflects the combination of humans (as a social system or a human system) in a changing environment (as an ecosystem). A socio-ecosystem-based approach addresses the environment – human links which contribute to resilience on climate change hazards. A socio-ecosystem is considered at different action levels: local, landscape, regional, countrywide, and global. At the local scale, communities enhance their resilience to the increasing risks on taking place their livelihood strategy, migration, and hazard preparedness. At the landscape and regional scales, governments issue adaptation policies and measures as coastal and beach protection, developing green cities, encouraging sustainable land use, applying integrated land use planning, and resilient housing.

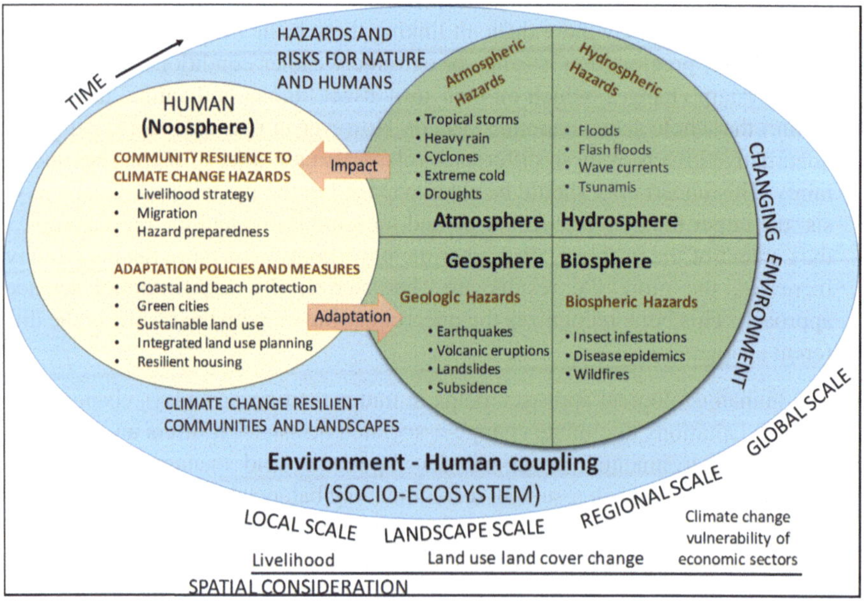

Fig. 1.3 Conceptual model for human ecology of climate change hazard

National and regional adaptation measures rely on community-based approaches, integrated land management strategies, and land conversion policies. Particularly in Vietnam, livelihood, land use land cover change, and climate change vulnerability of economic sectors are the most important aspects of socio-ecosystems and climate change resilient communities and landscapes.

Previous studies focused on specific aspects of human ecology of climate change hazards. In coastal areas and on the sea, the importance of changing human behavior to cope with the hazards was stressed. The effects of storms and future ecological responses on water quality in large estuaries in North Carolina were studied by Paerl et al. (2006). The result shows that the hydrologic load and the wind are significant drivers of the water quality in the estuaries. Adaptation to storms in coastal communities and small islands has been identified (Lane et al. 2013). All coastal communities including cities, towns, and villages situated along the shores of larger islands and continents are exposed to changing sea levels and increasing numbers and intensities of storms. The threats, vulnerabilities, and risks for local communities should be addressed by regional planning.

An important theme in applied human ecology deals with local responses to climate change. Butze (1983) and Carter (1987) reviewed responses to sea level change. Human ecology of storm surges, tornadoes, and marine hazards in coastal areas was analyzed (Aguirre et al. 1993; Chen et al. 1999; Li and Li 2013). Climate change contributes to migration and violent conflict (Reuveny 2007). Immigrants may cause conflicts in the local, hosting communities (Ding 2012). Consequently, the policy implications of climate change should receive priority in the reconciliation of conflicts. Using a simulation method to assess complex systems and predicting effects for a range of scenarios, the impacts of sea level rise and storms on the environment and people were studied for this book. The results advocate coastal flood management, erosion protection, and infrastructure design and planning, as necessary elements coping with sea level rise. Aalst et al. (2008) studied climate change adaptation using community risk assessments (CRAs). They concluded that risks could be reduced with proper adaptation policies. Granderson (2014) published on climate change risks and responses by communities using a cultural-political approach.

Recent decades witnessed growing attention for migration as an element of climate change adaptation. Migration flows in terms of duration, destination, and composition were studied. The issue should be subject to international, national, and local policies (IPCC 2007; Mortreux and Barnett 2009; Hugo 2011). During recent years, environmental refugees attracted the attention of climate change researchers (Farbotko and Lazrus 2012; Bettini 2013). Environmental refugees are displaced persons crossing borders, e.g., driven by climate change effects. The relationship between climate change and migration is complex and varies over time. Drought, desertification, and land degradation leading to food and water shortage are considered driving forces of local displacement as a result of the decline in agricultural yields. Extreme weather events are a main reason for affected people to leave their homes and move toward less hazardous areas. Sea level rise contributes to storms and floods. This offers a major threat to local people and households along the coast and in particular to the inhabitants of (small) islands (IPCC 2007). For example, the effects of climate change on migration affect people in the developing countries of Africa and Asia more and make them moving faster (Reuveny 2007). That is why migration is considered a strategy to climate change adaptation. In Vietnam, it is likely that both extreme weather events as storms, floods, and heat waves, but also sea level rise, temperature, and precipitation changes, will contribute to increasing migration (MONRE and UNDP 2010).

In the mountains, the slopes cover important parts of the agricultural area: over 8% of slopes are found among the 45% of the agricultural land worldwide (FAO 2012). The sustainability of fields on slopes shows environmental, economic, social, cultural, and governance aspects. An interdisciplinary approach is indicated to understand sustainable land use and local responses (Saint-Macary et al. 2010). Soil quality, agricultural land use, and natural hazards are closely related. Research on soil quality results in recommendations on cultivation practices, whereas the studies targeted to agricultural production highlight changes in land use policies. Acharya et al. (2008) and Graaff et al. (2008) introduced methods to protect soils and to reduce the impact of natural hazards in upland ecosystems. More efficient farming techniques and increasing the income of farmers under sustainable land use were recommended. Kelly and Huo (2013) assessed the efficiency of land conservation policies for farmers and recommended the contribution of farmers to land use policies. Regos et al. (2015) and Misbahuzzaman (2016) evaluated the effect of the agricultural land distribution and showed that land cover changes might result in both economic and environmental benefits. Victoria and Turner (2016) identified the changes in the livelihood of an ethnic group after the introduction of a new farming method on the slopes. They recommended the priorities for food security and agriculture development policies in the mountains.

Although climate change adaptation measures by governments are documented in official reports, records of the activities by locals are limited in the international and in the domestic literature. Information on adaptation measures by local communities and authorities can be collected using the Stakeholder Delphi technique (Bunting 2008; Le et al. 2015). A Delphi survey allows assessing the state of the problem, the trends, and the impacts on society when applied to climate change (Rowe and Wright 1999). Lemieux and Scott (2011) use a Delphi identifying and assessing climate change adaptation measures in Ontario, Canada. The desirability and feasibility of adaptation options were identified and assessed by a panel of protected area experts. Based on a multiple criterion decision-facilitation matrix, the research allowed identifying the most suitable options for a climate change action plan and ranking them by priority. In South Korea, fuzzy multi-criteria were used to assess the climate change-driven flood risk. Proxy variables were identified; their relative importance was weighted using a Delphi survey. The results allowed to assess vulnerability and to design adaptation strategies (Jun et al. 2013).

Box 1.1 Local Perception: Delphi Survey and Human Ecology of Climate Change Hazards

The Delphi survey contributed to early warning actions for climate change effects. In combination with the analytic hierarchy process (AHP) and dynamic fuzzy cognitive maps, Delphi enabled to define long-term planning options. The combination not only allows studying adaptation effects (Biloslavo and Dolinšek 2010), and determining early climate warning scenarios (Biloslavo and Grebenc 2012), but is also useful for multi-criteria decision-making (Kim and Chung 2013). Choi et al. (2012) studied climate change adaptation plans of four river basins in Korea. The results indicate that the watersheds became more vulnerable over time. Yoon et al. (2013) used a Delphi approach to assess the vulnerability of agriculture in Korea for global climate change. Exposure to particular climate conditions, sensitivity, and adaptation capacity of the community were evaluated. The sensitivity of the arable sector was compared with those in other regions during the period 2000–2020. Jung et al. (2014) used a Delphi technique searching for consensus among 50 water experts. The results showed the spatial distribution of the risks and provide a guide for investment and flood risk management in 229 localities. Lemieux et al. (2014) used Delphi surveying and identifying climate change vulnerability in watersheds to support planning and policy. The multi-sectorial impacts of climate changes encouraged a wide variety of experts and decision-makers to cooperate. In such complex situations, Delphi allows generating consensus among them on the essential elements for planning and public policy. In practice, this made it possible to come up with proposals to integrate climate change effects in planning and policy. Moreover, recognizing the relationship between climate change and water resources contamination resulted from a Delphi study (Kirezieva et al. 2015). The study identified the long-term community response strategies and pressures. Its outcome advocated that future adaptation options should systematically favor the adaptation capacity.

Delphi survey was applied in a number of studies on natural hazard mitigation. Luo et al. (2013) used a combination of Delphi and fuzzy AHP to analyze coastal erosion along the Fujian coast of Southeast China. Hazard, exposure, and coping capacity were mapped using a coastal erosion risk index. The risks subsequently were ranked and used to identify priorities for mitigation policies. Orencio and Fujii (2013) developed a disaster-resilience index for coastal communities using a combination of AHP and Delphi techniques. Decision-makers in the Philippines participating in a Delphi survey reached a consensus showing environmental and natural resource management, sustainable livelihoods, social protection, and planning are the most important factors defining resilience. Imaduddina and Subagyo (2014) used a Delphi survey to identify vulnerability factors influencing floods and sea level rise along the Surabaya coast of Indonesia. The results defined different risk levels and vulnerable zones by overlaying the weighted sum of the relevant factors.

1.1.3 Climate Change Hazards in Vietnam

Vietnam is one of the countries worldwide most prone to climate change associated with natural hazards (IPCC 2007, 2014). The country has 62 provinces and 7 geographic regions (Fig. 1.4). Both coastal areas and mountainous areas over Vietnam nowadays face a range of hazards. Coastal communities and populations living in the low delta areas are increasingly vulnerable to tropical storms, whereas mountainous communities and populations living in the upland are mostly affected by heavy rain-induced natural hazards as landslides, flooding, and flash floods (Dasgupta et al. 2009; Nguyen et al. 2011; Le et al. 2012; MONRE 2016).

During the period 1958–2014, the annual average temperature in Vietnam increased by 0.6°C (MONRE 2016). However, the trend was not temporally and spatially uniform: during the winter the temperature increased faster than in the summer, and the North tended to warm up faster than the South. The inland temperature increased faster than the coastal. Maximum and minimum temperatures increased significantly, with the highest rate of 1 °C over a 10-year period. Days warmer than 35 °C increased all over the country, at a rate of 2–3 days per 10 years in several regions. The last few decades saw several record high temperatures: the maximum temperature reached 42.7 °C in 2015. However, in the northwest, the number of hot days declined. Vietnam also witnessed an increase of the overall trend in the rainfall pattern, but the northern climate zones experienced less rain, while the southern ones got more (MONRE 2016).

The Vietnam Institute of Meteorology, Hydrology and Climate Change (IMHEN) was assigned by the Ministry of Natural Resources and Environment to take lead on updating the climate change and sea level rise scenarios for Vietnam and to cooperate with domestic and international research agencies to develop and update detailed climate change scenarios for Vietnam. The climate change scenarios version 2016 are based on the observed hydrometeorological data and sea level data updated to 2014; topographic data updated till March 2016; the most recent methodology of the 5th Assessment Report (ARS) of the Intergovernmental Panel on Climate Change (IPCC); global and regional climate models with high resolution; and dynamical downscaling methodology combined with statistical method for bias correction of model outputs. The effects of four scenarios were calculated: a low emission scenario (RCP 2.6), a low to medium emission scenario (RCP 4.5), a medium emission scenario (RCP 6.0), and a high emission scenario (RCP 8.5) (MONRE 2016). RCP 4.5 and RCP 8.5 are recommended officially as the initial orientation in assessing impacts of climate change and sea level rise in particular. They provide a basis to develop an action plan coping with these effects in Vietnam.

According to the RCP 4.5 scenario, the *average annual temperature* in Vietnam increased by 0.6–0.8 °C during the twenty-first century. By 2050, the temperature is expected to increase by 1.3–1.7 °C. More specifically in the Northern Vietnam, an increase by 1.6–1.7 °C is expected, while the corresponding figure for the North Central Region is 1.5–1.6 °C. In Southern Vietnam, the temperature is expected to increase by 1.3–1.4 °C. By the end of the century, the expectations are 1.9–2.4 °C in

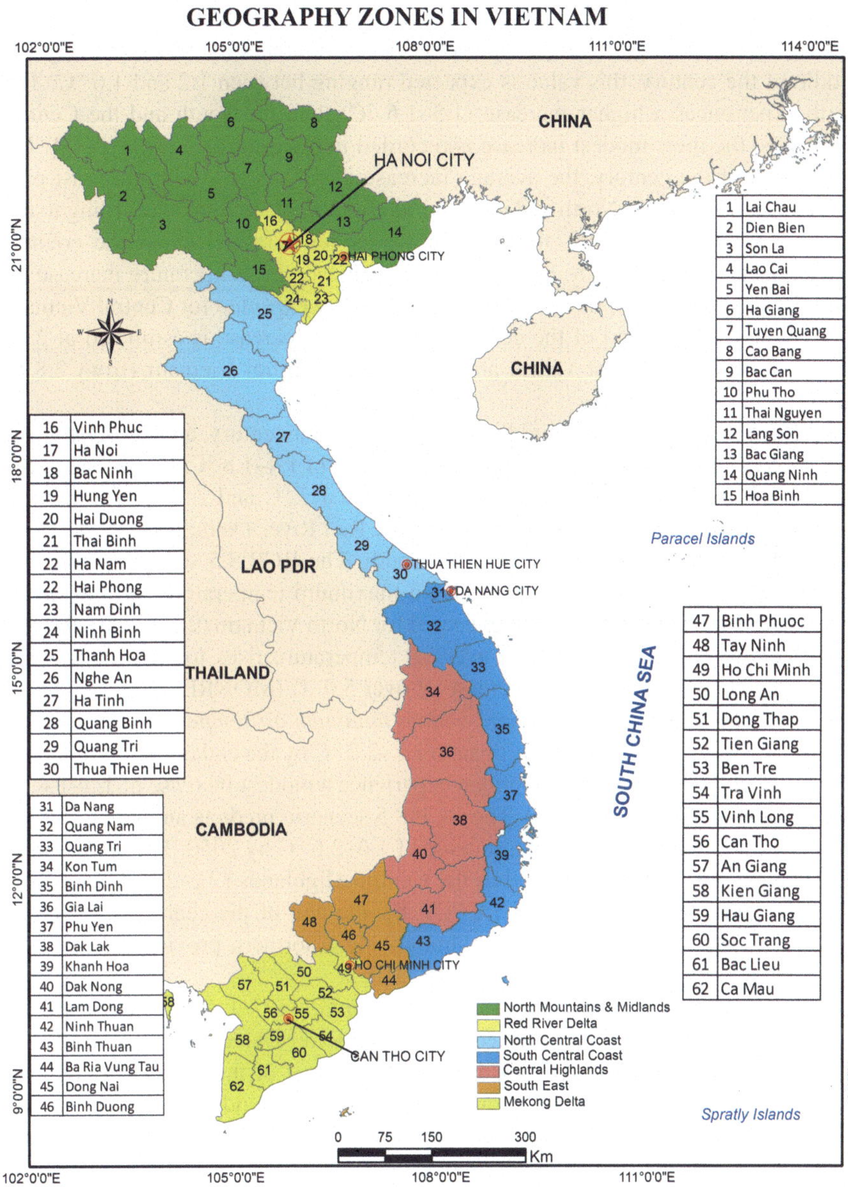

Fig. 1.4 Map of geographic regions and provinces in Vietnam

the North and 1.7–1.9 °C in the South. According to the RCP 8.5 scenario, by 2000, the annual average temperature in the country increased 0.8–1.1 °C. By 2050, the expected rise is between 1.8 and 2.3 °C. The northern region will be more affected (2.0–2.3 °C) than the South (1.8–1.9 °C). By the end of the century, the temperature in the North might increase by 3.3–4.0 °C and in the South by 3.0–3.5°C (MONRE 2016).

For the beginning of the twenty-first century, the RCP 4.5 scenario shows an increase in the *average winter temperature* across Vietnam of 0.6–0.8 °C. By the middle of the century, this value is expected ranging between 1.2 and 1.6 °C. The North experiences a higher increase (1.5–1.6 °C) than the South and the Central Highlands; the most modest increase is recorded in the Central coast (1.2–1.4 °C). By the end of the century, the average increase is 1.5–2.2 °C, with the most pronounced changes in the North and the Central areas of the country. According to the RCP 8.5 scenario, by 2000, the average winter temperature across the country increased on average by 0.8–1.2 °C. By 2050, an average temperature increase by 1.8–2.2 °C is forecasted; the most modest increase is expected for Central Vietnam (1.6–1.9 °C). By the end of the century, the average increase is estimated at 2.8–3.8 °C. Again, the lowest values are expected in Central Vietnam (from 2.8 to 3.2 °C).

RCP 4.5 forecasts, by the middle of the twenty-first century, an *average annual maximum air temperature* increase all over Vietnam of 1.4–1.8 °C. By the end of the century, the increase is expected ranging between 1.7 °C and 2.7 °C. The highest increase is expected for the North-East and the Red River Delta; the lowest values are for the South-Central region and the South. The RCP 8.5 scenario values by 2050 are an increase of the average annual maximum temperature nationwide by 1.6–2.4 °C. The highest increase is expected for North Vietnam (2.6 °C). By the end of the century, the average annual average temperature rises to 3.0–4.8 °C. The North of Vietnam might face an increase of over 5.0 °C (MONRE 2016).

The RCP 4.5 forecasts an *average annual minimum air temperature* increase of 1.4–1.6 °C by 2050. The figure will rise to 1.8–2.2 °C by the end of the century. The South-Central coast and South Vietnam experience a modest increase of 1.3–1.4 °C by 2050 and 1.6–1.8 °C by 2100. The RCP 8.5 scenario predicts an increase of the average minimum temperature nationwide of 1.6–2.6 °C by 2050. The temperature will change most in the North and in the Central Highlands (2.2–2.6 °C). In other areas the increase is gentler (1.6–1.8 °C). By the end of the century, an average increase of 3.0–4.0 °C is expected, although some northern provinces will likely face higher temperature changes (MONRE 2016).

RCP 4.5 foresees an increase in the *annual rainfall* in Vietnam, on average from 5% to 15% by 2050. Coastal provinces in the Red River Delta and the Central coast might face an increase of more than 20%. By the end of the century, the annual change in rainfall shows the same distribution as in the middle of the century, but selected regions will experience more than 20% of extra rain. RCP 8.5 starts from an increased annual rainfall all over country in 2000 (3–10%). By the middle of the century, the same trend as by RCP 4.5 is expected. By the end of the century, the highest increase of 20% is for the North, the Central Highlands, and part of the South (MONRE 2016).

Location and topography in Vietnam make the country prone to different types of natural disasters including tropical storms, floods, inundation, drought, desertification, salt intrusion, landslides, and earthquakes. About 59% of Vietnam's total land area and 71% of its population are vulnerable to cyclones and floods. From 1989 to 2009, Vietnam lost about 7286 million USD by natural disasters.

Storms and floods are the two main natural hazards which occur frequently and cause severe damages to lives, assets, and properties of locals. An estimated 80–90% of Vietnam's population is potentially directly affected by storms (CCFSC 2001). During the period 1954–2006, in total 380 storms and tropical depressions hit Vietnam, of which 31% in the North, 36% in the Northern Central part, and 33% in the South-Central and the South area. The top ten of the most expensive disasters in Vietnam shows that seven of them affected Central Vietnam. The post-disaster damage assessment and the reporting system tend to underestimate the economic value of the damage (CCFSC 2010). Vietnam will face even higher bills in the future as result of the increasing density of assets at risk, along the coast and in the lowlands (WB 2010).

Particularly in the lowlands, sea level rise averages approximately 2.5–3.0 mm per year during the past decades (Nguyen et al. 2011). More recent scenarios show that by the beginning of this century, the sea level rise in Vietnam is above the global average. The levels in the southern and eastern parts of the South China Sea are considerably above these in the other regions and in particular above these along the northern coast. By 2050 the RCP 4.5 estimates an average increase of 22 cm (14–32 cm) and by the end of the century of about 53 cm (32–76 cm). More drastic scenarios (RCP 8.5) foresee average increases of 25 cm (17–35 cm) in 2050 and 73 cm (49–103 cm) by 2100 (MONRE 2016). Currently no study is available indicating how much coastal land will be submerged neither estimating the number of Vietnam's over 3000 islands that will be submerged.

Storms and their consequences as floods are linked with other effects of climate changes: the intensity of the floods increases with increasingly higher sea levels; both storms and floods add to coastal erosion. The combination of tropical storms and sea level rise results in the destruction of houses and other infrastructure and triggers landslides which destroy habitats. The agricultural and densely populated lowland behind the dunes is most vulnerable to floods, which affect the communities. Moreover, the storm-climate changes interphase is a mutually enforcing system. Climate changes not only affect storms, but also, e.g., sea level rise results in a situation where the next storm builds on a larger column of water (IPCC 2007, 2014) (Fig. 1.5).

Over three quarters of Vietnam's territory however is covered by mountains. Mountain environments are affected by weathering, heavy rains, land collapses, and landslides; some of them are typical for weather events of the tropics (Sikor and Truong 2002; Schad et al. 2012; Delisle and Turner 2016). The variation in the landscapes results in considerable differences among the regions. Slopes in Vietnam offer opportunities for agriculture and forestry. Mountains and hills are the prevailing habitats of ethnic minorities. They mainly depend on agriculture using traditional techniques and means such as slash and burn. Most of the methods are rudimentary, outdated, and not supported by nowadays agroscience. As a result, the yields are limited and deeply influenced by natural conditions, which affect the upland farming systems and are a cause of environmental degradation. A great diversity of farming systems exists in the Vietnamese northern mountains, which

Fig. 1.5 A coastal landscape in the North of Vietnam. (Source: An Thinh Nguyen 2002)

results in complex terrains and a variety of climates and landscapes. Overall, the climate is typical for a tropical monsoon region with a huge potential for both forestry and agriculture. In the Vietnamese northern mountains, the Tay is the most numerous minority population (Le et al. 2012). Their indigenous knowledge and skills are the basis of agriculture on the slopes. In the northern mountains, the changes in climate and farm land are main drivers of land use change (Delisle and Turner 2016). Agricultural land in the deltas declines as a result of urbanization and industrialization, whereas the land on the slopes is increasingly under pressure by agriculture (Le and Tuan 2004; Saint-Macary et al. 2010). The effects offer an increasing challenge of contemporary pressures for the local sustainability in the mountains (Fig. 1.6).

An application of the human ecological approach to understand climate change impacts, situation, and local responses in both lowland and upland areas of Vietnam is needed. The likely effects of these hazards are merely described in comprehensive official reports (ISPONRE 2009; MONRE 2016); however measured data based on field experience and on local observations are rare in the international literature. Moreover, although the Vietnamese government enforces four national strategies on climate change, green growth, sustainable development, and environmental protection during the period 2010–2020 and a vision to 2030, specific solutions applicable to local districts, communes, and villages are scarce.

Fig. 1.6 A mountainous landscape in the North of Vietnam. (Source: An Thinh Nguyen 2006)

1.2 Methodology

1.2.1 Integration of Time and Space

1.2.1.1 Remote Sensing, GIS, and Spatial Modeling

Remote sensing is defined as "the collection of data about an object from a distance, the measurement of object properties on the earth surface using data acquired from aircraft and satellites" (Schowengerdt 1997). Remote sensing allows measuring objects from a distance, rather than in situ. The advancing remote sensors allowed a more effective monitoring of the Earth's environment. Often, these sensors are remote from the object of interest using helicopters, planes, and satellites. Sensors measure an object's electromagnetic energy reflected by and radiating from its surface. A remotely sensed system has seven components including energy source or illumination; radiation and the atmosphere; interaction with the target; energy; transmission, reception, and processing; interpretation and analysis; and application. Two types of remotely sensed systems exist: *passive remote sensing* which uses sensors that measure radiation naturally reflected or emitted from the ground or the atmosphere and *active remote sensing* techniques which use an artificial source of radiation.

Remote sensing has been applied to a wide array of fields, such as environmental assessment and monitoring, global change detection and monitoring, agriculture, nonrenewable resources and renewable natural resources exploration, meteorology, mapping, military surveillance and reconnaissance, and the news media. Geographers use techniques of remote sensing to monitor and measure phenomena in the Earth's lithosphere, biosphere, hydrosphere, and atmosphere. In human ecological studies, the remote sensing approach supports landscape change. Remote sensed data is collected at different spatial and temporal scales, offering a powerful tool to study processes. The availability of images captured by satellites, aircrafts, drones, or other observation systems combined with a multispectral reflectance allows a wide range of applications of these techniques to landscape change studies, such as boundary analysis (coastline change, coastal erosion), interpreting landscape complexity, and LULCC analysis.

Box 1.2 Object-Oriented Image Analysis for LULCC in the Ky Anh Coast and the Van Chan Mountain of Vietnam During 1993–2013

The object-oriented image analysis method considers both the spectral information of the image and shape and texture and contextual and semantic information. Its process includes two main steps:

- *Segmentation:* The image is segmented to describe meaningful objects. The object size is adjusted by the "scale" parameter which has to be defined prior to the segmentation process. The segmentation algorithm is a region-merging technique. It starts considering each pixel as a separate object. The number of image objects recognized depends on the scale defined by the user. The larger the scale, the fewer the number of identified image objects. For case studies on Ky Anh and Van Chan of Vietnam, both the images of 1993 and 2003 (LANDSAT TM) have been segmented at scale of 20. The color/shape ratio was set to 0.1/0.9, and compactness/sharpness ratio was 0.5/0.5 for the images. The images of 2013 (LANDSAT OLI) have been segmented at scale of 10. The color/shape was set to 0.1/0.9, and compactness/sharpness was set to 0.5/0.5.
- *Classification:* A rule-based approach was developed to classify built-up, forest, paddy land, and water using a nearest neighbor configuration for detailed classification. The nearest neighbor classifier assigns one of the four classes based on minimum distance measurements. The nearest neighbor is similar to supervised classification; training samples are needed prior to the classification.

Spatially distributed data are recorded at a variety of spatial and temporal scales. These data, particularly those derived from remotely sensed images, may be more efficiently stored and more effectively analyzed using a geographic information system (GIS). GIS is described as "an organized collection of computer hardware, software, geographic data and personnel designed to efficiently capture, store, update, manipulate, analyze, and display all forms of geographically referenced information" (Environmental Systems Research Institute, USA). Using GIS requires knowledge of GIS ports, peripheral devices, maps, image processing systems, GIS software, and networking and communication software. GIS is applied widely in most of the fields and has become a well-known tool all over the world. Parallel advancements in technology allowed the growth of GIS applications, from high-quality cartography to land use planning, natural resource management, environmental assessment and planning, tax mapping, ecological research, emergency vehicle dispatching, demographic research, utilities, and business applications. Particularly GIS-based decision supports are developed for land use planning in both lowlands and uplands worldwide (Buhlmann et al. 2010).

Remote sensing and GIS are combined to study shoreline change (Chen and Rau 1998; Boak and Turner 2005; Avinash et al. 2010; Pritam and Prasenjit 2010). These methods were used to detect shorelines and simulate shoreline change along the Ky Anh coastal area. An overall flowchart of the experimental method, which is shown in Fig. 1.7, consists of three steps including image pre-processing, coastline detection, and correcting effects of tidal levels.

- *Image pre-processing:* Once satellite images are geo-referenced, they are subset by using the study area boundary. The quality of subset images is then enhanced to highlight the shoreline by the edge filtering technique.
- *Coastline detection:* The coastline is the area of contact between river (sea) and land. As the sea level varies, this boundary varies, and therefore it is difficult to exactly identify its position (Boak and Turner 2005). The extraction of the coastline from a satellite image depends on the waterline, the tidal level, and several other parameters. Variations in the coastline position include long-term, cyclical, random effects due to specific events. The coastlines can be interpreted and digitized visually. In addition, the images can be classified into land and sea by identifying a threshold value for a single spectral band. An edge detecting filter or segmentation algorithm can be used to determine the shorelines from the images. Since the reflectance of water is nearly equal to zero in reflective infrared bands, and the reflectance of the majority of land cover is greater than water, coastlines can be extracted from a single-band image. For instance, band 5 of LANDSAT TM or ETM images can separate land and water features, but in the transition zone between land and water, this is influenced by mixed pixels and moisture regimes between land and water. If the reflectance values are sliced to two discrete zones, they indicate water (low values) and land (higher values) (Alesheikh and Nouri 2007). Since threshold values are not global, it is difficult to obtain accurate results with this method.
- *Correction effects for tides:* The definition of the coastline also considers variations along the shore. A coastline may also be considered over a slightly longer time scale, such as a tidal cycle, where the horizontal/vertical position of the shoreline varies between centimeters and tens of meters (or more), depending on the beach slope, tidal range, and prevailing wave/weather conditions (Boak and Turner 2005).

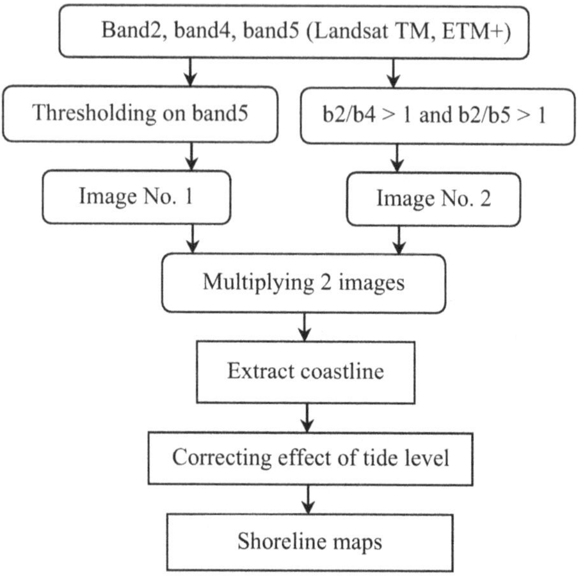

Fig. 1.7 Experimental flow chart. (Modified from Alesheikh and Nouri 2007)

Box 1.3 Shoreline Change Modeling for the Ky Anh Coastal Area

Tide level from 1973 to 2013 for each hour at the Hon Dau station and statistical data on the direction and speed of sea flows along the coastline was provided by the Institute of Marine Geology and Geophysics (IMGG), Vietnam. Satellite images used for extracting coastline include the LANDSAT MSS, TM, ETM+, and OLI images. Table 1.1 shows the details of the satellite images data and tidal height at their acquisition date.

Table 1.1 List of LANDSAT data and tidal height

Satellite images	Sensor	Parth/row	Date dd/mm/yyyy	Local time	Resolution (meters)	Tidal (meters)
LANDSAT 1	MSS	135/047	20 Jul 1973	9 h47′	80	0.14
LANDSAT 5	TM	126/047	07 Mar 1987	9 h39′	30	0.20
LANDSAT 5	TM	126/047	13 May 1994	9 h37′	30	0.27
LANDSAT 7	ETM+	126/047	12 Apr 2003	10 h06′	30	0.39
LANDSAT 8	OLI	126/047	08 Oct 2013	10 h19′	30	0.35

All LANDSAT images were geo-referenced to the UTM projection with a resolution of 30 meters. The band rationing method that is rationing between band 5 and 7 for LANDSAT MSS and also between band 2 and 5 for LANDSAT TM or ETM+ was used. Water bodies and land are separated clearly. Generally the ratio band 2/band 5 is greater than 1 for water and less than 1 for land in large areas of the coast (Pritam and Prasenjit 2010). The histogram of the rationing images was calculated, and the threshold was defined by changing the value step by step with intervals of 0.01 units until a clear difference between water bodies and land was obtained. The rationing image was sliced and converted to a shape file using the ENVI software version 5.2. Removing small objects and generating a shoreline map were performed using the ArcGIS software version 10.2.

In the Tri River Delta, tidal levels at all acquisition times of the images used varied from 0.14 meters to 0.39 meters. This means that tidal levels influence the extracted coastlines. As data are available on one coastline at a tidal level of 0.14 meters in 1973, this level was considered as the baseline. Other coastlines were adjusted based on this tidal reference level.

1.2.1.2 Networks Between Stakeholders

Local data is collected and compared from two main stakeholders as local households (farmers, aquaculturists, fish traders, or tourism managers) and local authorities. About a week prior to the field trip, the local authorities are invited preparing two reports: one on annual socioeconomic development in their commune or

Fig. 1.8 Site visit with the participation of research team, communal chair, and local farmers at the Ky Ninh beach of Ky Anh coast, March 2013. (Source: An Thinh Nguyen 2013)

district and another on natural hazards and the cost of their damage in Vietnamese dong (VND).[1] The reports were structured around issues such as the damage caused by natural hazards, their impact on livelihoods and the natural environment, existing socioeconomics and land use planning, and mitigation of and adaptation to the natural hazards. The reports were collected at the occasion of the field trip. They were introduced and put in context by the political and administration representatives of the communes and the district during a 45-min interview with the researchers. During the meeting, the policy lines of the local government on climate change coping and hazard prevention were assessed using the national strategies on climate change and green growth (2010–2020) as a benchmark. This document includes general options for adaptation to climate change hazards and green growth. The head of the local government scored the options as high, medium, or limited. After the meeting, local authorities guided the research team on a visit to the most damaged sites by recent hazards. Part of the household questionnaire allowed for comparison and independent validation of the reports by the authorities (Fig. 1.8).

1.2.2 Integration of Biophysical with Socioeconomic Data

1.2.2.1 Questionnaires

Questionnaires allowed inventorying the perception of the households about climate change and the impacts of natural hazards. Biophysical and socioeconomic data were collected, next to people's perception on the impacts on local livelihood capitals. All forms included two parts: the first part entails questions on climate change hazards and the other entailed questions on the specific effects of natural hazards. The questions were tested on their correct understanding in an independent test group, prior to their use in the study area. Questionnaires were completed by under- and postgraduate students in geography and economics of Vietnam National University, Hanoi, during a series of field trips during the period 2013–2017. All of them were trained to standardize the interviews with the local stakeholders. Completing a questionnaire took approximately 30–60 min (Fig. 1.9).

The most affected households were selected to complete the questionnaires. Because natural hazards damage part of the communes, the most affected villages were selected for the survey. Afterward random sampling was applied to identify the interviewed household representatives. This method allows a reasonable selection of the most affected families. Following the field trip, the results were tabulated in an excel frame, facilitating their treatment.

All questionnaire data were designed and processed to be qualified using a Likert scale. This allows grouping the data and ranking them at four levels. The perception of the local residents was inventoried using the original statement: "assess the natural hazards from the list below according to their impact use the Likert scale." The highest value on the scale reflects the most impacting natural hazard. The mean was used characterizing these data; mean values were expressed as the Likert wMean (weighted means). wMeans were used ranking natural hazards by their effects on the living conditions in the communities.

[1] The used conversion rate VND/$US is 22,727 (average rate in 2017).

Box 1.4 Data Collection Along the Ky Anh Coast Using Questionnaires
In Ky Anh, areas which are part of small villages along the coast inhabited by small populations were selected. Four separate questionnaires were completed by different family members (each person completed one questionnaire). Eighty-eight randomly selected respondents who represent the households in the most storm-affected villages of the three communes completed the questionnaires. The sampling fraction (the ratio between the number of respondents and total number of households most seriously damaged by the September 2013 storm in the selected villages) is 10.96%, 22.3%, and 5.33% for Ky Xuan, Ky Khang, and Ky Phu, respectively. The respondents were mainly (67–92%) farmers, and the overall sex ratio (1.15) indicates that slightly more adult males than females participated in the sample (Table 1.2). The sample provides a fair representation of the adult population in the villages.

Table 1.2 Number of questionnaires by type and characteristics of the interviewees in the Ky Anh coastal area

Questionnaires	Study area	Ky Xuan	Ky Khang	Ky Phu
Number of completed land use questionnaires (26 Q)	22	8	8	6
Number of completed green growth questionnaires (10 Q)	20	7	8	5
Number of completed livelihood quality questionnaires (15 Q)	26	9	9	8
Number of completed climate change questionnaires (20 Q)	20	9	6	5
Sampling fraction	88/890	33/301	31/139	24/450
Sex ratio of the interviews (male/female)	47/41	19/14	18/13	10/14
% of farmers answering the questionnaire	74	88	67	100
Age range of people answering the questionnaire	25–80	25–70	28–80	30–70

1.2.3 Stakeholder Delphi Survey

The Delphi survey was used to identify the impacts of climate change on the human society while establishing the Finish climate change policy. This provided an objective basis for policy recommendations (Wilenius and Tirkkonen 1997). Delphi survey also deals with potential adverse ecosystem impacts and the best available adaptation strategies for the future. Its results determined the existing state of ecosystems, climate change scenarios, and optimal adaptation strategies. This framework allowed identifying ecosystem management goals and management programs (Prato 2008).

The Delphi survey is widely used to gather data on business, environmental, social, economic, and other issues at present or in the future (Paliwoda 1983; Rowe

Fig. 1.9 Vietnamese students inventorying the perception of local farmers at their house in the Ky Xuan commune along the Ky Anh coast, August 2017. (Source: An Thinh Nguyen 2017)

and Wright 1999; Hasson et al. 2000; Linstone and Turoff 2002; Bunting 2008; Ballantyne et al. 2016). The approach proved most useful when science does not come up with precise answers (Gupta and Clarke 1996; Bunting 2008). At that moment Delphi allows identifying the consensus among the selected respondents (Hsu and Sandford 2007). More fundamental, Delphi structures a communication process allowing a group of individuals (panel members, respondents) to achieve consensus, eventually without face-to-face meetings (Paliwoda 1983). Different variations on the original Delphi technique exist. Among them are policy Delphi, Delphi conference, disaggregated policy Delphi, argument Delphi, and Stakeholder Delphi, to list just a few examples. The Stakeholder Delphi was selected because it not only allows identifying climate change hazards but also spots decisions on adaptation by both local stakeholders and authorities (Lund et al. 2014). The traditional Delphi process requires the involvement of experts only. They should be knowledgeable about the issue and include local authorities, planners, and decision-makers, next to "non-experts" (local residents) (Le et al. 2015). Data is collected from the respondents. Three different questionnaires were completed by the panel members aiming at reaching a high group consensus (Gupta and Clarke 1996). A general questionnaire was used during the preliminary stage (to identify objectives and define the problems the study should address), an open questionnaire in the first round (to rate the importance of the selected problems), and a closed questionnaire

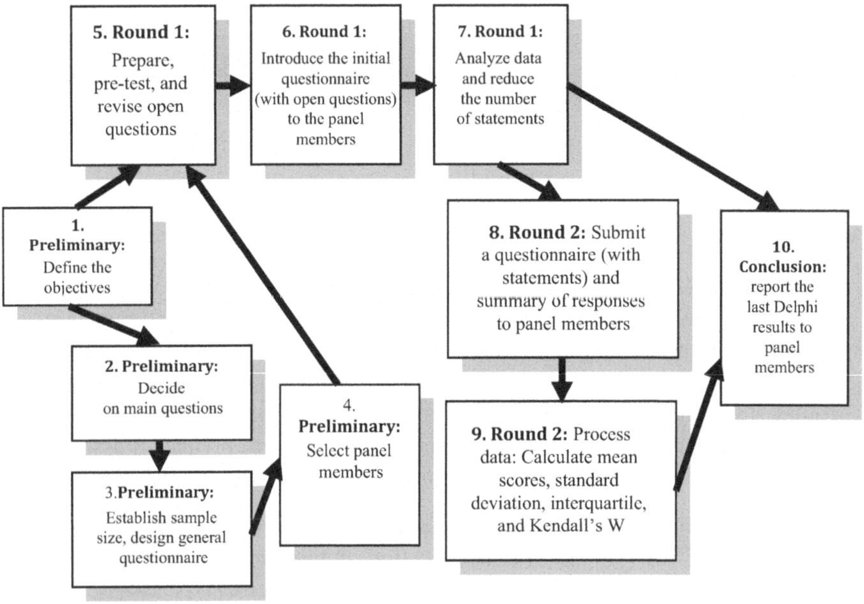

Fig. 1.10 Logical framework of the Stakeholder Delphi study

in the second round (to reassess the importance of the reduced statements based on the panel responses in the first round) (Rowe and Wright 1999). Each panelist completes his/her questionnaires independently from the others. After each round, the results are provided as a feedback to the respondents.

Figure 1.10 shows the rationale of this study in which the Delphi technique is applied in ten steps covering two rounds preceded by a preparatory phase:

- *Steps 1, 2, 3, and 4 (preliminary phase):* Once the objectives have been defined, a general questionnaire was developed introducing a general assessment of the impacts of the climate change-associated hazards on locals and the adaptation capacity. The number of statements was reduced using descriptive statistics on the collected data.
- *Steps 5, 6, and 7 (round 1):* These steps constitute the first round of the Delphi survey which was carried out during field trips. The steps include the preparation, pretesting, and revision of the open questions on statements, introducing the questionnaires to the panelists, and analyzing the data of which the results are used for feedback to the panel members. Finally the data provided a basis establishing a new closed questionnaire which is used during the second round. The first round questionnaire entails open questions based on two sources: a comprehensive literature review and information provided during the preliminary stage. Following this round a number of questions were modified, improving the clarity and the accessibility for the respondents. These changes allowed optimizing the questionnaires and adjusting the statements to the objectives. The result was a revised list of "questions" for the panel members.

Table 1.3 Interpreting the agreement and confidence associated with Kendall's W. (Schmidt 1997)

Value of Kendall's W	Agreement	Confidence in ranks
> 0.7–1.0	Very strong	Very high
> 0.5–0.7	Strong	High
> 0.3–0.5	Average	Average
> 0.1–0.3	Weak	Low
0.0–0.1	Very weak	No

- *Steps 8 and 9 (round 2):* These steps are part of the second round which was carried out during the next field trip. A questionnaire with the revised responses of the first round was sent to the same panel members who participated in round 1. This contributes to the coherence and the reliability of the results. The second round statements were refined using a selection of the most frequently mentioned alternatives in round 1. The panel members were invited to assess each statement in a Likert scale ranging from total disagreement to total agreement.

For each statement the mean score, the standard deviation, and the inter-quartile were calculated. This provides a first indication on the consensus or disagreement among the panel members. The reliability of the agreement is assessed using Kendall's coefficient of concordance (W) (Schmidt 1997) (Table 1.3).

- *Step 10 (conclusion):* Once the Delphi process was concluded, the results were disseminated to all panel members for their information. The final responses of the Delphi process were compiled.

Box 1.5 Selection of Panel Members for Delphi Surveys in the Ky Anh Coast and Van Chan Mountain of Vietnam

In the Ky Anh coastal area, 36 panel members in total were randomly selected from 4 stakeholder groups (sample size = 36): local government authorities (8), fishermen (8), fish traders (4), and farmers (16). All eight panelists of the authorities work at the administration of the communal department of resources and environmental management. The 28 panelists who are fishermen, fish traders, and farmers live in the villages which were most seriously affected by natural hazards: Nguyen Hue village (1 panel member), Xuan Thang (1), Le Loi (1), and Thang Loi (1) (Ky Xuan commune); Phu Hai (2) and Phu Loi (2) (Ky Phu); Trung Tien (2) and Trung Tan (2) (Ky Khang); Tan Thang (2), Tam Dong (2), Ban Hai (2), and Tam Hai (2) (Ky Ninh); Tan Thanh (1) and Hai Phong (1) (Ky Loi); Hong Hai (1) and Thang Loi (1) (Ky Phuong); and Minh Tien (1), Minh Thanh (1), Minh Quy (1), and Minh Duc (1) (Ky Nam).

In the Van Chan mountainous area, a total of 200 panel members were randomly selected from 4 local groups (sample size = 50 per community). All of the panelists are living in the villages which were most seriously affected by natural hazards: Son Thinh (50 Thai ethnicities), Thuong Bang La (50 Tay ethnicities), Suoi Giang (50 Hmong ethnicities), and Tran Phu (50 Kinh ethnicities).

1.2.4 *Analytical Complete (Unraveling Complexity)*

1.2.4.1 Indicators, Indexes, and Likert Scale

An indicator is a measured variable, which is considered revealing a characteristic, for example, environmental policy performance indicators, which were developed for environmental policy in the Netherlands (Adriaanse 1993). A list of indicators was built to calculate climate change vulnerability. Consider statements such as natural hazards, extent of the industrial production, potential aquaculture, and labor capacity indicating vulnerability of economic sectors under pressure of climate change. Vulnerability is "the degree to which a system is susceptible to, or unable to cope with the adverse effects of climate change, including climate variability and extremes" (IPCC 2001). Climate change vulnerability assessment is designed to quantify how communities adapt to climate change. Identifying the most vulnerable groups and areas necessitates establishing policies and adaptation measures. However it requires the conceptual framework, which is derived from the vulnerability concept in the Third Assessment Report of IPCC (2001). It is "a function of the character, magnitude, and rate of climate variation to which a system is exposed, its sensitivity, and its adaptation capacity" (IPCC 2001). For the Vietnamese Northern Central coast, climate change vulnerability (V) of economic sectors is calculated by the formula:

$$V = E + S + (1 - \mathrm{AC})$$

where V is vulnerability, E exposure; , S sensitivity, and AC adaptation capacity.

Exposure (*E*) expresses the nature and degree to which a system is exposed to significant climate variations. Sensitivity (*S*) indicates the degree to which a system is affected, either adversely or beneficially, by climate-related stimulation. Adaptation capacity (AC) points to the ability of a system to adjust to climate change, to moderate the potential damage from it, to take advantage of its opportunities, or to cope with its consequences (IPCC 2001). Six sectors of the Vietnamese Northern Central coast are selected to assess the vulnerability of climate change that include agriculture, aquaculture, forestry, industry, tourism, and population.

Measuring vulnerability necessitates multiple data, which was collected from the provincial statistical yearbooks. Climate indicators are based on climate monitoring data from 28 meteorological stations over the region during 34 years (1980 to 2013). Exposure indicators (E) include data from disaster reports of the Department of Agriculture and Rural Development and reports on natural disaster prevention and flood control of all provinces in the North Central Region. Data of sensitivity (S) and adaptation capacity (AC) are collected from statistical yearbooks during the period of 2000–2013, reports on socioeconomic performance during 2012–2013, and official report statistics on provincial agriculture and fishery.

Box 1.6 shows climate change vulnerability indicators for the Vietnamese Northern Central coast. Three groups of indicators are presented: exposure, sensitivity, and adaptation capacity. The first group includes 15 issues, and most of them are climate-related (14). The sensitivity consists of 11 subjects, and the last group on adaptation consists of 5 items. The selected subjects and indicators have a significant effect to a particular economic sector.

Box 1.6 List of Used Indicators to Assess Climate Change Vulnerability in Vietnamese Northern Central Coast

1. *Exposure (E)*: storm, flood, drought, annual average temperature, maximum temperature, minimum temperature, heat wave, damaging cold, yearly rainfall, rainfall variation in the rainy season, rainfall variation in dry season, rainy days, number of days with heavy rain, sea level rise, and agriculture-related indicators.
2. *Sensitivity (S)*: population-related indicators, natural hazard-related indicators, tourists, tourist facilities, type of tourism, natural reserves and national parks, extent of the industrial production, potential aquaculture, water surface area sensitive to flood, aquaculture, and forest sensibility indicators.
3. *Adaptation capacity (AC):* labor capacity, irrigation infrastructure, other infrastructure, economy, and policy mechanism.

An index is a composite variable. Indexes as the weighted mean value (*w*Mean), the standard deviation (*w*Std), and the consensus measure (CnS) were used characterizing questionnaire data, which were designed and processed by using a 5-point Likert scale. This method allows grouping the data and ranking them in five scales (Likert 1932; Duncan and Stenbeck 1987; Allen and Seaman 2007; Hartley 2014). The assessment scale used for the interpretation for *w*Mean is limited ($1.0 \leq w\text{Mean} < 1.8$), low ($1.8 \leq w\text{Mean} < 2.6$), medium ($2.6 \leq w\text{Mean} < 3.4$), high ($3.4 \leq w\text{Mean} < 4.2$), and very high ($4.2 \leq w\text{Mean} \leq 5.0$). For the consensus measure (CnS), an ordinal scale is used showing the degree of consensus (Tastle and Wierman 2005). The CnS provides values from 0 to 1 proportional to the increasing consensus among the respondents, in which 0 refers to total disagreement, while 1 stands for complete agreement.

The CnS is calculated using the equation:

$$CnS(X) = 1 + \sum_{i=1}^{n} p_i \log_2 \left(1 - \frac{|X_i - w\text{Mean}_X|}{d_X} \right)$$

where X_i is a particular variable of i, p_i the probability of the frequency associated with each X_i, d_x the width or span of X ($d_x = d_{\max} - d_{\min}$), X_i the particular Likert value, and $w\text{Mean}_X$ the weighted mean value of X_i.

1.2.4.2 Multiple-Criteria Decision Analysis (MCDA)

The MCDA-based geospatial approach evaluates explicitly multiple conflicting criteria in a geographical decision-making context. In this book, this approach is used for natural hazard zoning in uplands. The analytic hierarchy process (AHP) technique is selected for MCDA. AHP, which was developed by Saaty (1980), is a semi-quantitative method. It involves a matrix-based pair-wise comparison of the contribution of different factors for land sliding. AHP deals with quantifiable and tangible criteria and is applied to decision theory and conflict resolution. AHP is used to build a hierarchy of decision elements (factors) and comparisons between possible pairs in a matrix to give a weight to each element and a consistency ratio. The factor weight of each criterion is determined by a pair-wise comparison matrix (Table 1.4) (Saaty 1990, 1994, 2001).

To assess the consistency of the AHP results, a consistency ratio (CR) is used. It is obtained by comparing the consistency index (CI) with the average random consistency index (RI). The consistency ratio is defined as:

$$\mathrm{CR} = \frac{\mathrm{CI}}{\mathrm{RI}}$$

The average random consistency index (RI) is derived from a sample of randomly generated reciprocal matrices using the scales 1/9, 1/8,…., 8, and 9. The consistency index (CI) is calculated using the equation:

$$\mathrm{CI} = \frac{\lambda_{\max} - n}{n-1}$$

where $\lambda_{\max}$ is the maximum eigenvalue of the matrix and n is the order of the matrix.

Table 1.4 Scales for pair-wise comparisons

Scales	Degree of preferences	Descriptions
1	Equally	Two activities contribute equally to the objective
3	Moderately	Experience and judgment favor slightly to the moderately favor one activity over the other
5	Strongly	Experience and judgment favor strongly or essentially favor one activity over the other
7	Very strongly	An activity is strongly favored over another, and its dominance is shown in practice
9	Extremely	The evidence favoring one activity over another is affirmative of the highest degree possible
2, 4, 6, 8	Intermediate values	Used to represent compromises between the preferences in weights 1, 3, 5, 7, and 9

Source: Saaty (1990, 1994, 2001).

The consistency of the AHP is used to build a matrix that is checked by a consistency ratio, which depends on the number of parameters. The CR must be less than 0.1 to accept the calculated weights. The models with a CR above 0.1 are automatically rejected, while a CR less than 0.1 is as a rule acceptable (Saaty 1990, 1994, 2001).

Box 1.7 The MCDA-Based Geospatial Approach for Natural Hazard Zoning in the Van Chan Mountains

The AHP was used to produce hazard zone maps for the Van Chan mountains. The relevant thematic layers on causal factors were generated using field visit data and GIS. Each value of the Landslide Hazard Zone Index (LHZI) and the Flooding Hazard Zone Index (FHZI) for each considered pixel was calculated by multiplying each factor's weight with the class weight of each referred factor (for that pixel). This is written as:

$$\mathrm{L(F)HZI} = \sum_{i=j}^{n} \left(W_i R_j \right)$$

where L(F)HZI *is the required Landslide (floods) Hazard Zone Index of the given pixel and R_j and W_i are the class weights (or rating values) and the factor weight for factor i (derived using AHP).*

Factors as elevation, aspect, and slope were derived from the digital elevation model (DEM) with 100 meter contour. Lineament density and rock were derived from a geology map. Stream density, land use land cover, soil, and maximum precipitation were found on topology, land use land cover, soil, and rainfall maps published by the MONRE and MARD in 2015. The scale of the geographic maps was 1/50,000. All the collected data were converted to a raster grid with 100 m × 100 m cells to use AHP. The ArcGIS software was used to prepare thematic maps and layers.

1.2.5 Combination Methods

1.2.5.1 The DPSIR Framework Combined with Stakeholder Delphi

Since the Organization of Economic Cooperation and Development (OECD 1993) and the European Environment Agency (EEA 1995) developed the DPSIR framework, the model allows structuring, clarifying, and identifying solutions for environmental problems. This is in particular the case when DPSIR is combined with other methods as the analytic hierarchy process (AHP), ecosystem-based management (EBM), the outcome approach, or integrated coastal management (ICM)

(Gari et al. 2015). The results of the combination of DPSIR with Delphi provide a common view of the stakeholders on cause-effect relationships (Lund et al. 2014; Le et al. 2015). In this book the Stakeholder Delphi technique is combined with the DPSIR framework applied to the coast of Ky Anh.

The Delphi process uses 20 questions in each of both rounds. These questions were listed according to the DPSIR components:

- Two questions on the drivers (D): agriculture and nonagricultural sectors.
- Three questions on the pressures (P): migration, calamities, population growth, mining industry, processing industry, and agriculture.
- Five questions on the state (S): frequency of extreme weather events, intensity of storms, and trend of floods and droughts.
- Four questions on the impacts (I): changes in agricultural land and yield of crops.
- Six questions on responses (R): construction and upgrading dikes and irrigation systems.

Applying the DPSIR framework to establish Delphi questionnaires allows analyzing the relationships between the origins and consequences of climate change hazards. Moreover, the links between the DPSIR components help to understand the dynamics of local climate change hazards.

Box 1.8 Combining the Delphi Survey and the DPSIR Framework: A Literature Review

Worldwide, a few studies apply a combination of Delphi and the DPSIR framework. They deal with natural resources and environmental management, climate change adaptation, and hazard mitigation (Omann et al. 2009; Newton and Weichselgartner 2014; Gari et al. 2015; Lewison et al. 2016). The HighARCS (Highland Aquatic Resources Conservation and Sustainable Development) project used Stakeholder Delphi techniques to support participatory action in planning for sustainable aquatic resource use in the mountains of Shaoguan (China), West Bengal and Uttarakhand (India), and Son La and Quang Tri (Vietnam). A combination of the DPSIR and the STEP models was used to develop the questionnaires. Panel members were selected from different stakeholder groups including government authorities, captains of industry, fishermen, and farmers. Friedman's chi-square test and Kendall's W at a significance level of $p < 0.05$ were calculated. This provided information on the degree of consensus among the respondents. The results show that the Stakeholder Delphi supports decision-making on planning (Lund et al. 2014). Benitez-Capistros et al. (2014) used a combination of the Delphi technique and the DPSIR framework to assess environmental impacts and to measure ecosystem services on the Galapagos Islands (Ecuador). Eight impacts were identified including the introduction of alien species, biodiversity loss, land use change, biological resource loss, habitat fragmentation, landscape alterations, water basin overexploitation, and water quality degradation. The Delphi

(continued)

Box 1.8 (continued)
results provided an objective basis for decision-making on conservation management. A two-round Stakeholder Delphi, in combination with the DPSIR framework, was used to collect data and opinions of local communities and authorities on climate change impacts on agriculture in the Thai Binh province, Vietnam (Le et al. 2015). The results contributed to capacity building among the local communities on climate change policies both province- and nationwide.

1.2.5.2 Damage Valuation Combined with Trend and Cost Analysis

Damage valuation is the process of information collection, inventorying, and analysis of the negative impacts of natural hazards. The results help decision-makers acting on rescuing, emergency interventions, and prevention strategies. A combination of damage valuation, trend analysis, and cost analysis is applied to the coast of Ky Anh. Tropical storms and related hazards cause damage and physical losses and destroy the life of local people. Damage valuation of the effects of tropical storms entails five steps (Table 1.5):

1. *Damage data collection:* Data collection on damage on-site is based on documents, construction records, purchased equipment, and facilities during the storm and the restoration of the damage. This was assessed on a daily basis using official inventories and reports.
2. *Monetization:* The rough monetary figures were provided by the local authorities of the communes and those of the Ky Anh district. They were interviewed during the period 23–26 February 2014. The interview was guided by a questionnaire they completed in beforehand.
3. *Livelihood damage valuation:* The rough monetary data were put into perspective using the methodology of Chambers and Conway (1992). Livelihood was defined as a set of assets, involving access to drinking water, food, fodder, medicine, shelters, and cloths and the capacity to acquire these assets either individually or as a group (DFID 1999). Data and statistics of the damage were collected from the People's Committees of the district and the communes in Ky Anh. Based on the available data about compensation or replacement, the impacts on three livelihood capitals are calculated: the *natural capital* (destroyed soils and forests), the *physical capital* (damaged transport, shelter, electricity, information, and telecommunication), and the *financial capital* (loans and cash used to recover/repair assets of household after the storm). A lack of information in the communities explains why the human and the social capitals were not calculated. The data provided allow quantifying indicators of livelihood damage valuation. The same indicators have been used by the MARD Vietnam Central Committee for Flood and Storm Control (CCFSC 2006).
4. *Trend analysis using a time series analysis model:* The nonlinear ARIMA (autoregressive integrated moving average) model allows predicting damage valuation in the future (Kaushik and Singh 2008; Momani and Naill 2009;

Shamsnia et al. 2011; Mahsin et al. 2012; Bari et al. 2015). The method was used by Box and Jenkins (1976) and modified by Box et al. (1994). Because this model has been applied on economic prediction (Cui 2011), it is used to predict the future behavior of damage variables. Input data for the ARIMA model are damage figures during the period 2008–2013 (6 years). Yearly damage data series during 2008–2013 was calculated by monetization (data statistics by local authorities) and based on indicators used by the CCFSC (2006). These data were provided by the local authorities at the occasion of the documented interviews. The model estimates difference equations containing stochastic components and uses two approaches: autoregressive (AR) and moving averages (MA). ARIMA describes non-stationary time series based on a three-stage methodology including *model identification* (identifying the appropriate degree of differentiation), *estimation* (using the autocorrelation function ACF and the partial autocorrelation function PACF to find a tentative model and estimating the parameters of

Table 1.5 Research rationale and the coherence of the damage valuation applied to Ky Anh coast (Ha Tinh province, Vietnam)

Particular steps		Aims	Used methods	Comments/ limitations
1. Data collection	1.1. Household questionnaires	Perception of locals about the effects of tropical storms	Household questionnaire Descriptive statistics. Likert scales	The study is limited to the most affected households by tropical storms
	1.2. Sectorial impacts	Observations and opinions of local administrations Assess the impacts of tropical storms locally	In-depth interviews, filter data, and reports from local authorities	Data is used for livelihood damage valuation; trend and cost analysis
2. Damage valuation of tropical storms	2.1. Monetization	Assess the cost of damage based on monetization value	In-depth interview and questionnaire interviews with local authorities	N/A
	2.2. Livelihood damage valuation	Assess and compare impacts of storms and related natural hazards on different livelihood capitals during the period 2008–2013	Sustainable livelihood framework	Lack of information for the human and the social capitals
	2.3. Trend analysis	Findings about the damage value prediction for the period 2014–2019	Time series analysis	Time series of input data ranges for 6 years
	2.4. Cost analysis	Compare and assess forecasted damage taking into account two interest rates	Cost analysis	Benefit from storms was considered as negligible value

the model), and *validation* (the residual analysis and forecast). The damage by tropical storms and related hazards is predicted until 2019 (6-year forecast). This is a valid time interval for ARIMA, although the results should be interpreted with caution if outlying values appear.

5. *Cost analysis:* The cost analysis is based on the cost-benefit analysis (CBA) and aims at assessing the damage (cost) caused by tropical storms in the Ky Anh coast during the period 2014–2019. The CBA is widely used in Environmental Impact Assessment (EIA) to compare benefits and costs of projects for the environment. A CBA allows combining economic (cost, benefits of different scenarios) with environmental and resource benefits and costs (Tran et al. 2015; Söderqvist et al. 2015). Although benefits of storms have been referred in a number of works (Dasgupta et al. 2009; Shreve and Kelman 2014), the benefits from these hazards are negligible. Values are expressed as the net present value (NPV) of a time series of cash flows. The accumulated NPV during future years is the sum of the NPV of the livelihood capitals. The NPV results of the cost analysis provide estimates of the amount of money needed to support affected communes on hazard prevention and management.

1.3 The Challenge

Climate change-associated effects of extreme weather conditions are most pronounced in Vietnam. Table 1.6 indicates a human ecological approach to study effects of natural hazards on the nature and humans at national, regional, and local scales.

This approach has been optioned for because of its:

- *Multi- and interdisciplinary character:* All main sectors including agriculture, industry, business, tourism, and households are involved.
- *Wide range of methods:* Methods dovetailing in natural sciences as well as in human science disciplines are used. Geographical data are combined with perception studies.
- *Integration of data:* The multitude of studies used in this work results in data with a range of different characteristics. Data are integrated using models as the DPSIR framework and the combination of damage valuation with trend analysis and cost analysis. Stakeholder Delphi techniques are used obtaining consensus among different target groups providing a basis for recommendations.

The recommendations of this human ecological analysis of climate change hazards and their related impacts aim at contributing to more adaptation capacity and target both the stakeholders (farmers, aquaculturists, tourism managers) and decision-makers.

Table 1.6 360 degree overview of the human ecological aspects of climate change-associated natural hazards in the book chapters

Chapter	Scientific problem	Methodology	Studied scale
Chapter 1	Human ecology of climate change hazards	Systematic review	International and national scales (Vietnam)
Chapter 2	Climate change-associated hazards, impacts, and vulnerability	IPCC's climate change vulnerability concept	Regional scale (the Vietnamese Northern Central coast and Northern Mountainous region)
Chapter 3	Climate change hazards and migration	Spatial analyst; questionnaires; Likert scale and descriptive statistics	Local scale (the Ky Anh coast: Ky Xuan, Ky Khang, and Ky Phu communes)
Chapter 4	Gross costs and impact on local livelihood capitals of tropical storms	Networks between stakeholders; damage valuation combined with trend and cost analysis	Local scale (the Ky Anh coast)
Chapter 5	Climate change adaptation of local communities along heavily damaged coasts	Combination of Stakeholder Delphi and the DPSIR framework	Local scale (20 coastal villages of the Ky Anh coast)
Chapter 6	Impacts and damage of climate change hazards in the upland	MCDA; remote sensing, GIS, and modeling	Provincial and district scale (Yen Bai province, Van Chan district)
Chapter 7	Comparing local and immigrant household preparedness for natural hazards in the upland	Two round – Stakeholder Delphi	Local scale (Kinh, Tay, Thai, and Hmong villages of the Van Chan mountain)
Chapter 8	Building resilient landscapes and green cities along the coast and in the upland	Systematic analysis	Multi-scales (international, national, regional, and local)

1.4 Conclusion

This introductory chapter puts emphasis on a series of headlines of this book. First on its human ecological nature includes complex problems as natural hazards of which humans are both at the origin and the subject of the effects, can be studied using a human ecological approach, even when this is not evident methodologically, and characterized by uncertainties. These latter should be explained in a transparent way. Human ecology allows integrating different levels of analysis, characterized by their specific uncertainties. In this way, a quite complete picture of the problem under study appears. It combines data on biophysical aspects with among others health, economy, physiology, and social aspects. This is not only useful from a deeper understanding point of view but results in recommendation which can be applied by planners and discussion makers.

Second on the methods used consists of a combination of methods dovetailing in both natural and human sciences is used. Mainly the combination of geography with social and perception studies prevails when it is about the science of natural hazards. This combination leads to a wide range of methods of which these allowing integration of data and focusing on sustainable development are the most eminent.

In this way this chapter provides basic information backing the seven remaining chapters in this book.

Chapter 2
Climate Change-Associated Hazards, Impacts, and Vulnerability at Regional Level

Abstract The field research in this book is about human and natural replies to main climate change-associated effects as storms, floods, and landslides in two areas: the Ky Anh coastal district in the North Central Vietnam and the Van Chan mountainous district in the Northern Vietnam. While in the open literature ample attention is given to these climate change-associated effects in deltas and in the lowlands near the sea, less information is available about how people in the mountains experience and adapt to the changing mountainous climate.

Climate variation exists in a long country as Vietnam bordered by an over 3000-km-long coast. Therefore this chapter overviews the changing climate in the wider geographical context of both study areas: the 6 coastal provinces of North Central Vietnam and the 15 provinces of the Northern mountainous region. The first part of this regional approach describes core climate aspects as temperature and wet precipitation along the coast of Central Vietnam. This is followed by the phenomenology of storms, flash floods, droughts, and landslides. This section is concluded with an analysis of the vulnerability of the six main sectors in this area: agri- and aquaculture, forestry, industry, tourism, and the dense population living here. The second part of this chapter aims at a similar approach adapted to the mountains in the northern part of the country. Temperature changes and drought are analyzed, next to floods, flash floods, landslides, and periods of drought and (low) temperature extremes.

Keywords Climate change impacts · Climate change vulnerability · Natural hazards · Economic sectors · Regional scale · Vietnamese Northern Central coast

A. T. Nguyen, L. Hens, *Human Ecology of Climate Change Hazards in Vietnam*, Springer Climate, https://doi.org/10.1007/978-3-319-94917-8_2

2.1 Climate Change Hazard Impacts and Vulnerability Along the Vietnamese Northern Central Coast

2.1.1 The Northern Part of the Central Coast of Vietnam

The Vietnamese Northern Central coast is bordered by the South China Sea in the east (with islands and archipelagoes), while small and narrow plains, mountains, and hills are located in the west (Le et al. 2012). This region consists of six coastal provinces (Thanh Hoa, Nghe An, Ha Tinh, Quang Binh, Quang Tri, and Thua Thien Hue) and counts six provincial cities, including Thanh Hoa (Thanh Hoa), Vinh (Nghe An), Ha Tinh (Ha Tinh), Dong Hoi (Quang Binh), Dong Ha (Quang Tri), and Hue (Thua Thien Hue). Cities are located along the national road no. 1A (Fig. 2.1). Cities play an important role as the administrative, industrial, commercial, service, education, and training centers of the region.

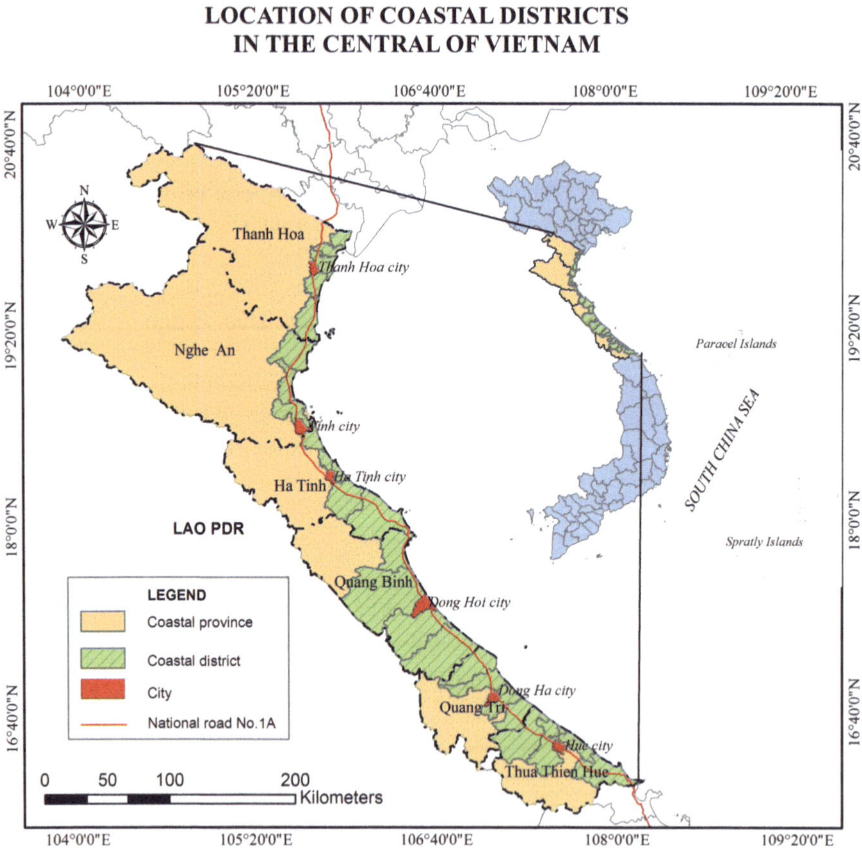

Fig. 2.1 Location of the provinces and provincial cities along the Vietnamese Northern Central coast

By 2013, the population of the Vietnamese Northern Central coast totals about 10.3 million persons accounting for 11.5% of the national population and ranking fifth among the seven economic regions of Vietnam. The Thanh Hoa province hosts the largest population: by 2013, it reached over 3.4 million inhabitants. The smallest province is Quang Tri with slightly more than 0.6 million inhabitants. The population density in 2013 of these central provinces is currently 193 people per square kilometer, which coincides with 0.7 times the average population density of the country (271 people per square kilometer). The Thanh Hoa province has the highest population density (312 people per square kilometer), which is 1.6 times the population density of the region and 1.2 times this of the country. The lowest population density is in Quang Binh (107 people per square kilometer) (GSO 2014).

The GDP of the region increased from 1.26 billion $US in 2000 to 11.28 billion $US in 2013. The share of the agriculture-forestry-fishery sector decreased from 40.5% (in 2000) to 21.6% (in 2013); the industry-construction sector on the other hand increased from 22.5% in 2000 to 34.9% in 2013, and the service sector increased from 37% to 43.5%. Overall, the development of the industry boosted in line with the industrialization and modernization of the country's economy. Although the value of industrial production in the region is still limited as compared to Vietnam as a whole, it increased from 2.4% in 2000 to 4.7% in 2013 (GSO 2014). The economic structure of the region changes; however the domestic activities are still most significant, while foreign investments are negligible.

These advanced areas of infrastructure, technical facilities, science, tax, and technology are located along the coast and therefore vulnerable to climate change impacts. The effects of storms and floods should be part of their impact assessments.

2.1.2 Climate Change

2.1.2.1 Average Air Temperature Scenario

- *Average annual temperature:* The average expected increase in annual air temperature (°C) along the Vietnamese Northern Central coast is summarized in Table 2.1. Values between brackets show the variation around the mean value with the lower limit of 10% and an upper one of 90%. For example, by the middle of the century, in the Thanh Hoa province, the average increase according to RCP 4.5 ranges between 1.1 and 2.3 °C, with a mean of 1.7 °C. The average annual temperatures tend to increase by 0.6 °C to 3.7 °C during the period 2020–2100 as compared to 1986–2005. For each period and scenario, the temperature raises most in Thanh Hoa and Nghe An and the least in the Thua Thien Hue. The rate of increase rises fast between 2000 and 2050 according to scenario RCP 4.5, while in scenario RCP 8.5, the rate accelerates by the end of the century (MONRE 2016).

Table 2.1 Average increase in air temperature in the Vietnamese Northern Central coast as compared to the baseline period 1986–2005 for two scenarios (°C)

	RCP 4.5			RCP 8.5		
Provinces	2016–2035	2046–2065	2080–2099	2016–2035	2046–2065	2080–2099
1. *Average annual temperature*						
Thanh Hoa	0.7 (0.3–1.1)	1.6 (1.1–2.3)	2.2 (1.6–3.2)	1.0 (0.6–1.5)	2.1 (1.4–3.2)	3.7 (2.9–5.2)
Nghe An	0.7 (0.3–1.1)	1.6 (1.1–2.2)	2.2 (1.5–3.1)	1.0 (0.6–1.5)	2.0 (1.4–3.1)	3.7 (2.9–5.2)
Ha Tinh	0.6 (0.3–1.0)	1.5 (1.0–2.1)	2.0 (1.4–2.9)	0.9 (0.6–1.3)	1.9 (1.3–2.8)	3.5 (2.8–4.8)
Quang Binh	0.6 (0.3–1.1)	1.5 (1.0–2.1)	2.0 (1.5–2.8)	0.9 (0.6–1.2)	1.9 (1.3–2.8)	3.3 (2.7–4.7)
Quang Tri	0.6 (0.4–1.2)	1.4 (1.0–2.0)	1.9 (1.3–2.8)	0.9 (0.6–1.2)	1.9 (1.3–2.7)	3.3 (2.6–4.6)
Thua Thien Hue	0.7 (0.4–1.1)	1.4 (0.9–2.0)	1.9 (1.3–2.7)	0.8 (0.6–1.2)	1.9 (1.3–2.6)	3.3 (2.6–4.5)
2. *Average temperatures in winter*						
Thanh Hoa	0.6 (0.3–1.1)	1.4 (0.9–2.0)	1.8 (1.1–2.6)	1.0 (0.6–1.4)	1.9 (1.3–2.7)	3.2 (2.3–4.3)
Nghe An	0.7 (0.3–1.1)	1.4 (0.9–1.9)	1.7 (1.1–2.4)	1.0 (0.6–1.4)	1.8 (1.2–2.6)	3.1 (2.2–4.2)
Ha Tinh	0.6 (0.3–1.0)	1.3 (0.7–1.8)	1.6 (1.0–2.1)	0.9 (0.6–1.2)	1.7 (1.2–2.4)	2.8 (2.0–3.7)
Quang Binh	0.6 (0.2–1.0)	1.2 (0.8–1.7)	1.5 (0.9–2.0)	0.9 (0.6–1.2)	1.7 (1.2–2.3)	2.8 (2.0–3.6)
Quang Tri	0.7 (0.3–1.1)	1.2 (0.8–1.6)	1.5 (1.0–2.0)	0.8 (0.5–1.1)	1.7 (1.2–2.2)	2.8 (2.2–3.5)
Thua Thien Hue	0.7 (0.3–1.1)	1.2 (0.7–1.6)	1.4 (0.9–2.0)	0.8 (0.5–1.1)	1.6 (1.2–2.1)	2.8 (2.2–3.4)
3. *Average temperature in spring*						
Thanh Hoa	0.6 (0.0–1.2)	1.4 (0.8–2.0)	2.1 (1.3–3.2)	0.9 (0.5–1.4)	1.9 (1.0–3.1)	3.4 (2.3–4.9)
Nghe An	0.7 (0.1–1.3)	1.4 (0.8–2.0)	2.1 (1.3–3.1)	1.0 (0.5–1.5)	1.9 (1.0–3.1)	3.4 (2.2–5.0)
Ha Tinh	0.6 (0.1–1.2)	1.3 (0.7–1.9)	2.0 (1.2–2.9)	0.9 (0.5–1.3)	1.8 (0.9–2.8)	3.2 (2.0–4.5)
Quang Binh	0.6 (0.2–1.2)	1.3 (0.8–1.8)	2.0 (1.2–2.8)	0.9 (0.4–1.2)	1.8 (0.9–2.8)	3.2 (2.1–4.5)
Quang Tri	0.7 (0.2–1.2)	1.3 (0.8–1.8)	2.0 (1.3–2.8)	0.9 (0.5–1.2)	1.8 (1.0–2.7)	3.2 (2.2–4.6)
Thua Thien Hue	0.7 (0.2–1.2)	1.4 (0.7–1.9)	2.0 (1.2–2.8)	0.9 (0.5–1.2)	1.9 (1.0–2.7)	3.2 (2.2–4.5)
4. *Average temperature in summer*						
Thanh Hoa	0.8 (0.4–1.2)	1.9 (1.3–2.9)	2.7 (1.9–3.7)	1.0 (0.5–1.5)	2.4 (1.5–3.6)	4.1 (3.1–5.9)

(continued)

Table 2.1 (continued)

	RCP 4.5			RCP 8.5		
Provinces	2016–2035	2046–2065	2080–2099	2016–2035	2046–2065	2080–2099
Nghe An	0.8 (0.3–1.3)	1.9 (1.3–3.0)	2.7 (1.9–3.7)	1.0 (0.5–1.6)	2.4 (1.4–3.7)	4.2 (3.2–6.0)
Ha Tinh	0.8 (0.4–1.3)	1.9 (1.2–3.0)	2.6 (1.8–3.6)	1.0 (0.5–1.5)	2.3 (1.4–3.6)	4.1 (3.2–5.7)
Quang Binh	0.8 (0.4–1.3)	1.8 (1.2–2.9)	2.5 (1.7–3.5)	0.9 (0.5–1.5)	2.2 (1.4–3.4)	4.1 (3.0–5.8)
Quang Tri	0.8 (0.3–1.3)	1.8 (1.2–2.7)	2.3 (1.7–3.3)	0.9 (0.5–1.5)	2.2 (1.4–3.2)	3.8 (3.0–5.5)
Thua Thien Hue	0.7 (0.3–1.3)	1.7 (1.2–2.7)	2.4 (1.7–3.3)	0.9 (0.6–1.4)	2.2 (1.4–3.2)	3.8 (3.0–5.4)
5. *Average temperature in autumn*						
Thanh Hoa	0.6 (0.2–1.1)	1.7 (1.1–2.5)	2.2 (1.4–3.2)	1.0 (0.5–1.6)	2.2 (1.4–3.3)	3.9 (2.9–5.4)
Nghe An	0.6 (0.2–1.1)	1.6 (1.1–2.4)	2.1 (1.3–3.2)	1.0 (0.4–1.6)	2.1 (1.4–3.3)	3.9 (2.8–5.5)
Ha Tinh	0.6 (0.3–1.1)	1.5 (1.0–2.2)	2.0 (1.2–2.9)	0.8 (0.4–1.4)	2.0 (1.3–3.0)	3.6 (2.7–5.0)
Quang Binh	0.6 (0.2–1.2)	1.5 (0.9–2.1)	1.9 (1.2–2.8)	0.8 (0.4–1.3)	1.9 (1.3–2.9)	3.5 (2.7–4.8)
Quang Tri	0.6 (0.3–1.2)	1.3 (0.9–2.1)	1.8 (1.2–2.8)	0.8 (0.4–1.3)	1.8 (1.3–2.8)	3.4 (2.7–4.6)
Thua Thien Hue	0.6 (0.3–1.1)	1.4 (0.9–2.1)	1.8 (1.2–2.7)	0.8 (0.4–1.3)	1.9 (1.2–2.8)	3.4 (2.6–4.6)

Source: MONRE (2016)

- *Average temperatures in winter (December to February):* According to RCP 4.5, the winter temperature increases by 0.6 to 1.8 on average. The highest increase is expected in Thanh Hoa and the lowest in Thua Thien Hue. The increase is faster by the beginning of the twenty-first century and tends to slow down toward the end of the century. According to RCP 8.5, the average winter temperature increases from 0.8 to 3.2 °C. The temperature will increase the most in Thanh Hoa and the least in Thua Thien Hue. However, in contrast to RCP 4.5, the increase speeds up by the end of the century. For example, in Ha Tinh, the increase rises from 0.9 to 1.7 °C but jumps to 2.8 °C in the late 2000s.
- *Average temperature in spring (March to May):* According to RCP 4.5, the average increase in temperature during spring time is 0.6–0.7 °C by 2000. By the middle of the twenty-first century, the increase rises to 1.3 or 1.4 °C. This will result in an extra 2.0 or 2.1 °C by 2050. Thanh Hoa will experience the highest temperature increase (2.1 °C), while the Quang Binh, Quang Tri, and Thua Thien Hue provinces will have the lowest increase of 2.0 °C by the end of the century. According to RCP 8.5, by 2100, all provinces along the Northern Central Coast face a spring temperature increase of 0.9 °C, except in Nghe An, where the temperature will rise by 1.0 °C. In contrast to the RCP 4.5 results, the temperature

Table 2.2 Average change in annual precipitation (%) compared to the baseline period 1986–2005 (lower limit of 20% and upper limit of 80%) in the Vietnamese Northern Central coast provinces according to two core scenarios

	RCP 4.5			RCP 8.5		
Province	2016–2035	2046–2065	2080–2099	2016–2035	2046–2065	2080–2099
Thanh Hoa	10.1 (3.7–16.8)	17.6 (11.5–23.6)	21.3 (14.2–29.0)	13.8 (8.5–19.0)	18.6 (13.0–24.5)	25.5 (19.9–31.2)
Nghe An	10.2 (2.4–17.7)	16.8 (10.6–23.1)	18.1 (10.3–26.3)	16.6 (7.7–24.5)	21.6 (14.1–28.5)	26.4 (18.8–33.6)
Ha Tinh	11.3 (6.0–16.6)	16.3 (8.5–24.4)	13.0 (3.4–22.6)	12.9 (6.8–18.9)	14.1 (8.9–19.0	17.4 (10.6–24.4)
Quang Binh	10.1 (3.5–16.5)	12.6 (3.8–22.0)	10.9 (0.0–21.4)	10.8 (4.0–17.4	14.1 (8.2–19.6)	12.1 (5.5–19.0)
Quang Tri	11.4 (2.9–20.0)	16.6 (7.5–26.2)	20.1 (9.8–31.3)	16.5 (9.9–22.8)	16.8 (10.7–22.6)	16.4 (8.2–24.2)
Thua Thien Hue	17.0 (10.4–23.6)	22.5 (10.7–34.3)	26.2 (15.4–38.1)	16.5 (9.0–23.3)	18.6 (12.9–23.9)	21.2 (13.8–28.2)

Source: MONRE (2016)

will rise faster toward the end of the century (on average by 0.9–3.2 °C). The increase is most pronounced in Thanh Hoa and Nghe An where temperature rises by 3.4 °C.

- *Average temperature in summer (June to August):* According to RCP 4.5, the summer temperature will increase the least in Quang Tri, reaching a value of 2.3 °C by the end of century, while Thanh Hoa and Nghe An face the highest increase (2.7 °C). For the period 2016 to 2065, all provinces are expected to experience almost the same increase in temperature. RCP 8.5 forecasts that Quang Tri and Thua Thien Hue will experience the most modest increase (3.8 °C), while the temperature will change most in Nghe An (4.2 °C). The rate of increase doubles by 2100 as compared to the start of the century.
- *Average temperature in autumn (September to November):* RCP 4.5 shows that the Northern part of the Central Coastal provinces experienced an average increase of 0.6 °C by 2000. By the middle of the century, the temperature increase is most modest in Quang Tri (1.3 °C) and the most pronounced in Thanh Hoa (2.2 °C). According to RCP 8.5, the temperature increases substantially by the end of the century, with 3.9 °C in the Thanh Hoa and Nghe An provinces.

2.1.2.2 Precipitation Scenario

Along the Vietnamese Northern Central coast, according to RCP 4.5, Thua Thien Hue will experience most increase in rain (17% by 2000 and 26.2% by 2100). All provinces will face more precipitation except Quang Binh where a decrease of 12.6–10.9% is expected. Also in Thanh Hoa, the rain increases substantially over the century (from 10.1% to 21.3%). RCP 8.5 shows similar expected effects as RCP

4.5; however, most rain will fall in Nghe An (16.6%). The precipitation will increase by 26.4% in 2100. Both Quang Tri and Quang Binh will witness a decrease in the rate of change by the end of the century (Table 2.2).

2.1.3 Natural Hazards

2.1.3.1 Storms

Landfalls of storms usually accompanied by high tide and heavy rain result in long periods of rain and floods. The flood season in Central Vietnam lasts from June to October. Along the rivers between Quang Binh and Binh Thuan, the flood season lasts from September to December. The Central region has short and steep rivers with high debits. Dike systems in this region are relatively low or incomplete. 8-meter-high floods not only occur along the main streams but also spread over the floodplains (Le et al. 2012).

Between 1951 and 2010, 285 tropical storms were recorded all over Vietnam (on average 4.75 tropical storms per year); during the period from 2011 to 2013, the coastal area of Vietnam faced 38 tropical storms (Matsumoto and Shoji 2003). However, during the more recent two and a half decades (1996–2005), only four storms were recorded or on average 0.4 storms per year (ISPONRE 2009). Particularly along the Vietnamese Northern Central coast, storms in this area increased between the 1950s and the 1980s but subsequently decreased during the 1990s. The peak month of the storm-induced landfalls shifted from August in the 1950s to November in the 1990s, and considerable uncertainty exists about the expected frequency of storms during the remaining part of this century. Storms moved southward in recent years, though it is widely expected that because of the increasing temperature, the North will face more storms in the near future. Also the intensity of the storms is expected to increase, resulting in more wind and more intense precipitation (CCFSC 2001; IPCC 2007). In particular, more intense storms, representing in more threats to people's lives, livelihoods, infrastructure, and agriculture, are forecasted.

In 2009, storm Ketsana affected provinces along the Vietnamese Central coast, killing 163 people and causing over 600 million $USD of damage (CCSFC 2010).

In 2010, storms and other natural hazards killed or caused missing 173 people. 168 others were injured in October 2010 (GSO 2014).

In 2012, the South China Sea faced 12 storms, of which 4 directly affected Central coast.

In 2013, Central Vietnam was hit directly by consecutive storms. The Wutip storm in September 2013 damaged over 1000 houses (Vietnam NCHMF 2013). Over 70,000 people in vulnerable areas were moved to shelters along the central coastline (Al Jazeera America, accessed November 22, 2013). In November 2013, the Haiyan storm forced over 800,000 people to evacuate. Storm Nari in November 2013 destroyed about 12,000 houses in 7 central provinces (The Weather Channel, accessed November 22, 2013).

In 2016, six tropical depressions and ten storms affected the Vietnamese Central coast. Six storms and one tropical depression directly impacted the land.

In September 2017, Central Vietnam was hit by the Doksuri storm. Over 100,000 people were evacuated, 4 people died, and 10 were injured. The storm Doksuri caused heavy rains and floods all over the provinces in the Vietnamese Northern Central coast. Thousands of houses were damaged or destroyed. More than 50,000 houses in Ha Tinh, Quang Binh, Quang Tri, and Thua Thien Hue provinces were damaged. Quang Binh People's Committee reported that about 200,000 houses were flooded or submerged, 5000 lost their roofs and 20 collapsed (updated news on Vietnamnet website, accessed on 15 September 2017).

2.1.3.2 Floods, Flash Floods, Droughts, and Landslides

Besides storms, the Vietnamese Northern Central coast faced other natural hazards such as floods, flash floods, droughts, landslides, and coastal erosion.

By 1996, over 2000 square kilometers of the Vietnamese coast was estimated to be at risk for annual floods. Flood damage is expected to worsen if the daily rainfall increases by 12–19%. The floods of September 2002 affected the Huong Son, Huong Khe, and Vu Quang districts of the Ha Tinh province. Consequently, 53 persons died, while 111 people were injured. The floods and flash floods of September 2011 in the Thanh Hoa and the Nghe An provinces caused six deaths (CCFSC 2014).

Drought intensified as a result of the increased variation in rainfall and evaporation (3% along the coast and 8% inland by 2070). The effect is triggered by rising temperatures (MONRE 2016).

Landslides in the Northern Central coast are often triggered by heavy rains and storms, resulting in large amounts of sliding material downhill. Riverbank erosion is widely spread in this region, in particular during the rainy season. The lower part of the rivers is severely affected. Coastal erosion goes up to 10 meters annually, which worsens with the sea level rise of the recent years. The most intensive erosion occurred in the Ky Anh (Ha Tinh province) in the recent years.

2.1.4 Climate Change Vulnerability

2.1.4.1 Vulnerability of Agriculture

The agriculture of the Vietnamese Northern Central coast is based on annual industrial crops such as peanuts, straw, sugarcane, mulberry, and pineapple (Le et al. 2012). Recently, tea, rubber, coffee, pepper, and coco were introduced. Peanuts cover an area of over 40,000 hectares and are mainly grown in the Nghe An, Ha Tinh, and Thanh Hoa provinces. Over 35,000 hectares of sugarcane are planted.

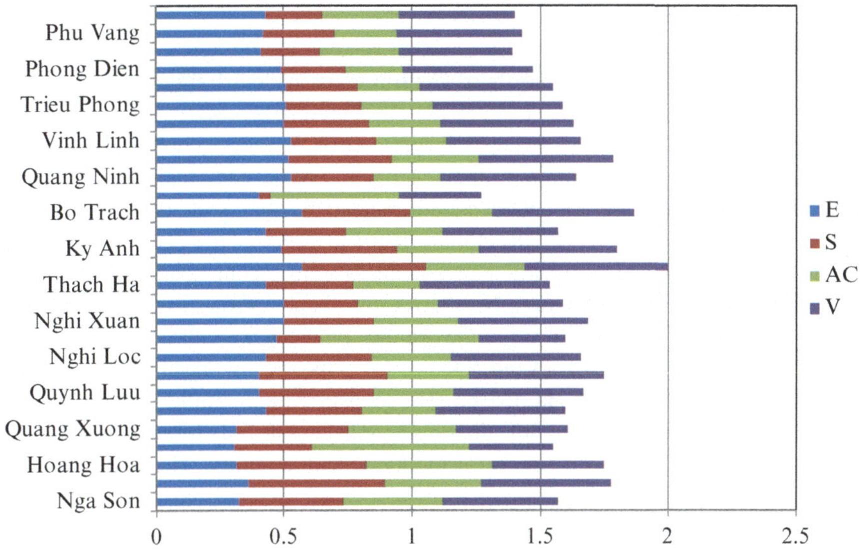

Fig. 2.2 Climate change vulnerability assessment of the agriculture by district of the Vietnamese Northern Central coast. Horizontal axis indicates value of exposure (E), sensitivity (S), adaptation capacity (AC), and vulnerability (V)

Perennial crops as pepper are grown in the Quang Binh and Quang Tri provinces; coffee, rubber, and tea are grown in Nghe An, Quang Tri, and Thanh Hoa. Fruit trees are common in the Tan Ky and Nghia Dan (Nghe An) and Van Du and Ha Trung (Thanh Hoa) districts. Livestock consists of cattle, pigs, goats and deer in Nghe An, Ha Tinh, and ducks in Thanh Hoa (GSO 2014). The vulnerability of agriculture in the districts depends on extreme climatic events. Most districts in the Ha Tinh, Quang Binh, and Quang Tri provinces have a high exposure because they suffer storms, floods, and drought. Districts with a high exposure index show also a high vulnerability. For example, the Cam Xuyen district (Ha Tinh province) with the highest exposure in the region (0.57) represents the highest vulnerability (0.56). This underlines that the agriculture in the region with traditional methods mainly depends on the weather conditions.

The overall vulnerability index of agriculture ranges between 0.32 and 0.56 for all coastal districts along the Vietnamese Northern Central coast. This corresponds with a moderate to high vulnerability. No district experiences a low vulnerability to climate change. The value is lowest in the Dong Hoi city (0.32), while the highest value is attributed to the Cam Xuyen and Bo Trach districts (0.56). Most of the districts (17/28, 11 of them belong to Ha Tinh, Quang Binh, and Quang Tri provinces) show a high vulnerability to climate change, whereas 11 other districts show a moderate vulnerability (Fig. 2.2).

2.1.4.2 Vulnerability of Aquaculture

Provinces of the Vietnamese Northern Central coast have a long coastline, many estuaries, lagoons, and bays (Le et al. 2012). Aquaculture is promoted and gradually became a leading economic sector. Shrimp, crab, seahorse, holothurians, and *Gracilaria asiatica* are the main products. Aquaculture farmers, including both fish and crustaceans, are water-dependent and influenced the quality of coastal resources. Higher temperatures and more droughts affect the yields. This is ongoing as the yields of the spring crops declined drastically during recent years (GSO 2014). Aquaculture along the Vietnamese Northern Central coast shows high vulnerability to climate change: the vulnerability index ranges between 0.33 and 0.73. The highest value (0.73) is for the Gio Linh (Quang Tri province), while the lowest value (0.33) applies to the Thach Ha district (Ha Tinh province). Aquaculture shows a high vulnerability in majority of the districts (25/28), while only three districts (Sam Son, Cua Lo, and Thach Ha) report a moderate vulnerability. The exposure and sensitivity index of aquaculture are the highest of all sectors considered. The districts in the Quang Tri and Thua Thien Hue provinces show the highest vulnerability because of its high sensitivity (Fig. 2.3).

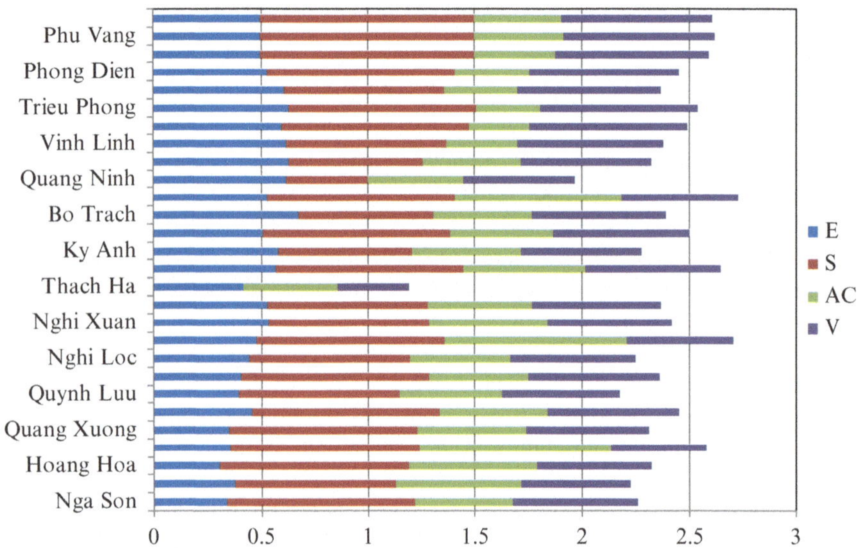

Fig. 2.3 Climate change vulnerability assessment for aquaculture by district of the Vietnamese Northern Central coast. Horizontal axis indicates value of exposure (E), sensitivity (S), adaptation capacity (AC), and vulnerability (V)

2.1.4.3 Vulnerability of Forestry

The Vietnamese Northern Central coast entails the second largest forest area in Vietnam (2.46 million hectares). Nghe An and Quang Tri provinces have the highest wood yield in this region and Vietnam (GSO 2014). Authorities paid attention to greening hills, afforestation of coastal sandbanks, and green belts in urban centers and industrial parks. Forest trees include pine, acacia, eucalyptus, casuarina, manglietia, conifers, and local trees in the coastal area. Mangroves spread along the coast, prevent floods, and mitigate the damage of storms and heavy rains. Primary forests are only found on the high mountains in the Laos-Vietnamese borderlands. Other natural forests are limited in protected areas (Le et al. 2012). Most forested areas were planted on the lower mountains and hills.

The forests show a moderate vulnerability by climate change (the vulnerability index ranges between 0.29 and 0.56). The value is lowest in Dong Hoi city (Quang Binh province), and the highest value is found in the Bo Trach district of the Quang Binh province. Twenty-four districts show a medium vulnerability. Only four districts experience a high index: this applied to the districts Cam Xuyen (Ha Tinh province), Bo Trach, Quang Ninh (Quang Binh province), and Vinh Linh (Quang Tri province). The districts with the highest vulnerability are Cam Xuyen (0.51) and Bo Trach (0.56). Exposure is high in the Cam Xuyen (0.6) and Bo Trach (055) districts. Forestry shows a similar, moderate exposure as agriculture, but the sector has a higher adaptation capacity; consequently the vulnerability degree is lower (Fig. 2.4).

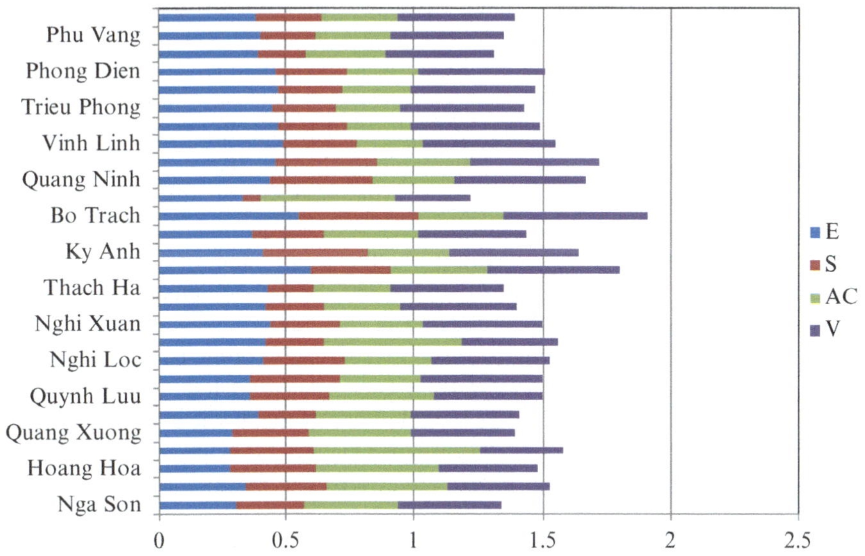

Fig. 2.4 Climate change vulnerability assessment for forestry by district of the Vietnamese Northern Central coast. Horizontal axis indicates value of exposure (E), sensitivity (S), adaptation capacity (AC), and vulnerability (V)

In recent decades, the provincial government invested more financial resources and issued policies to protect and manage the forests and the mangroves. In particular planting mangrove forest along coasts with high risk of storm and sea level rise has been addressed. The districts in the Ha Tinh province where most forests were planted show the lowest vulnerability. In contrast, districts which did not benefit of the planting policy show the highest vulnerability.

2.1.4.4 Vulnerability of Industry

In the Vietnamese Northern Central coast, during 2000–2012, the average growth rate of industry is over 14% per year (GSO 2014). Since 2000, more significant coastal economic zones (EZ) have been established boosting the industrial development of the region. The Nghi Son EZ (Thanh Hoa) spreads over 18,612 hectares and is the most important economic zone of the Thanh Hoa province and the Nam Thanh – Bac Nghe area. The Southeast Nghe An EZ (Nghe An) covers 18,826.47 hectares. It is a hub for international trade in the Northern Central Coast, an important gateway to Central Vietnam, to the Laos PDR and northeastern Thailand. The Vung Ang EZ (Ha Tinh) extends over 22,781 hectares, combining industry, trade, and tourism. The Hon La EZ (Quang Binh) is 10,000 hectares and focuses on the marine economy, the production of building materials, electricity, and tourism. The Chan May-Lang Co EZ (Thua Thien Hue) covers 27,108 hectares and has a deepwater seaport. The Southeast Quang Tri EZ was established in 2010 (Le et al. 2012). These economic zones attract deepwater industry, establish a new momentum for the socioeconomic organization, and are part of territorial networks linking industrial zones.

The majority of economic zones locate near the shoreline. This makes them vulnerable to climate change hazards. However, industry is less affected as compared to agriculture, forestry, and aquaculture. The industrial zones resist the effects of natural disasters easier. This explains that the industry is moderately vulnerable to climate change: this relates to the moderate qualification of exposure, sensitivity, and adaptation capacity of most of the districts. The high vulnerability in seven districts is related with the high exposure. Industrial plants in new areas which do not offer solid constructions and modern equipment are more at risk from natural hazards than other areas. The vulnerability index is lowest (0.33) in the Sam Son (Thanh Hoa province) and highest (0.57) in the Bo Trach (Quang Binh province). Seven districts (25%) along the coast show a high vulnerability of industry to climate change. In the remaining 75% of the districts, the industry is considered moderately vulnerable. Overall the Quang Tri province experiences the highest vulnerability, while the index is the lowest in the Thanh Hoa province (Fig. 2.5).

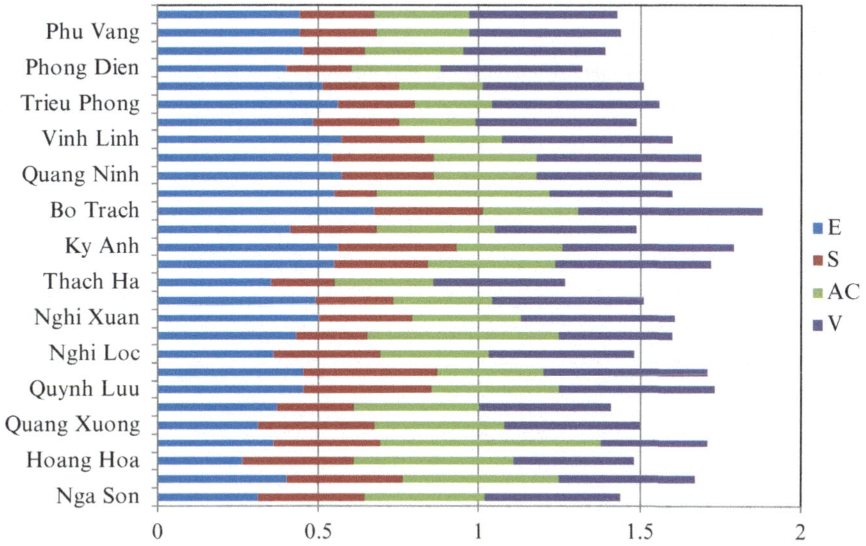

Fig. 2.5 Climate change vulnerability assessment for industry by district of the Vietnamese Northern Central coast. Horizontal axis indicates value of exposure (E), sensitivity (S), adaptation capacity (AC), and vulnerability (V)

2.1.4.5 Vulnerability of the Population

The Vietnamese Northern Central coast shows its uneven distribution of the population, which reflects a difference between the eastern coastal plains and the western hilly and mountainous areas (Le et al. 2012). Most of the population is located along the national road no. 1A and in the eastern coastal plain, which accounts for over 70% of the population and which is more dense than the national average. Hilly and mountainous areas in the West account for 60% of the area, but only 30% of the people live in this region. Consequently, the average density in the western mountains of the country is only about 10–50 people per square kilometer (GSO 2014).

Natural hazards damage habitats of locals in hilly and mountainous areas as well as coastal areas, while storms and flash floods impact both uplands and lowlands. These latter are affected by a combination of storm, floods, sea level rise, and coastal erosion. This explains why the region has a moderate to high vulnerability of the population to climatic change. The calculated vulnerability index ranges between 0.30 and 0.64. The lowest index (0.30) is found in the Cua Lo (Nghe An province), while the highest value (0.64) is in the Bo Trach (Quang Binh province). Seventeen districts experience a high vulnerability; the remaining 11 districts show a moderate vulnerability (Fig. 2.6). All coastal districts of the Quang Tri province with a high vulnerability show also high values for exposure and sensitivity. This is not surprising as people tend to look for the best conditions to live and avoid the places that adversely affect their lives.

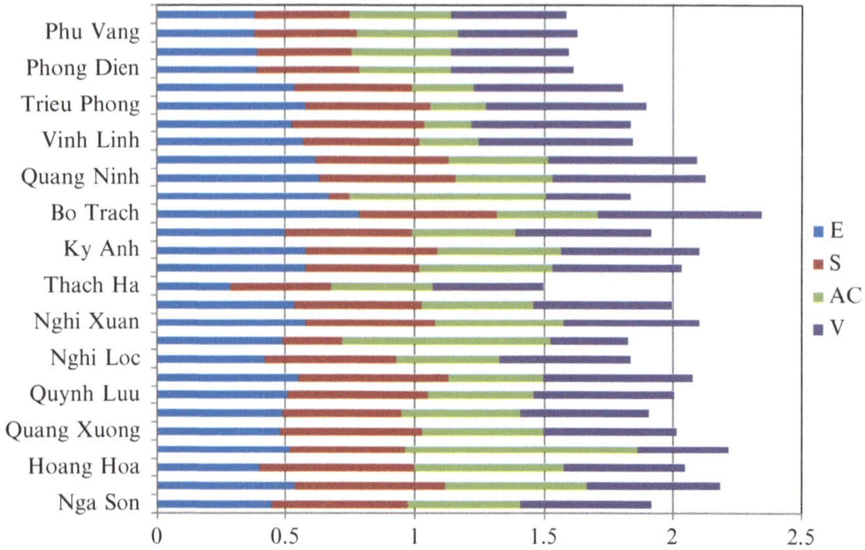

Fig. 2.6 Climate change vulnerability assessment for population distribution by district of the Vietnamese Northern Central coast. Horizontal axis indicates value of exposure (E), sensitivity (S), adaptation capacity (AC), and vulnerability (V)

2.1.4.6 Vulnerability of Tourism

In the Vietnamese Northern Central coast, tourism becomes a leading economic sector as the region has plenty natural and cultural tourism attractions:

- The most significant sandy beaches locate in Sam Son (Thanh Hoa), Cua Lo (Nghe An), Nhat Le (Quang Binh), Cua Tung (Quang Tri), Thuan An, and Lang Co (Thua Thien Hue).
- Natural landscapes along the Huong river, the Tam Giang lagoon, the Hai bridge (Thua Thien Hue), the Thien Cam mountain, and the Ke Go lake (Ha Tinh).
- National parks include Ben En (Thanh Hoa), Pu Huong, Pu Mat (Nghe An), Vu Quang, Ke Go (Ha Tinh), Phong Nha-Ke Bang (Quang Binh), and Bach Ma (Hue).
- Cultural and architectural monuments entail the Hue citadel, Vinh Moc tunnel, Truong Son cemetery, Con Tien island, and Quang Tri citadel. Festivals in Lam Kinh (Thanh Hoa) entail among others the temple festival of Cuong (Nghe An) and the festival of Hon Chen (Thua Thien Hue).
- The UNESCO World Heritage sites include the Ho citadel (Thanh Hoa), the Phong Nha-Ke Bang (Quang Binh), and the Hue Imperial Citadel (Thua Thien Hue).

Currently, the government invests in developing marine tourism, ecotourism, speleo-tourism, and heritage tourism along the Vietnamese Northern Central coast. However, climate change affects the cultural monuments. Also the water supply in

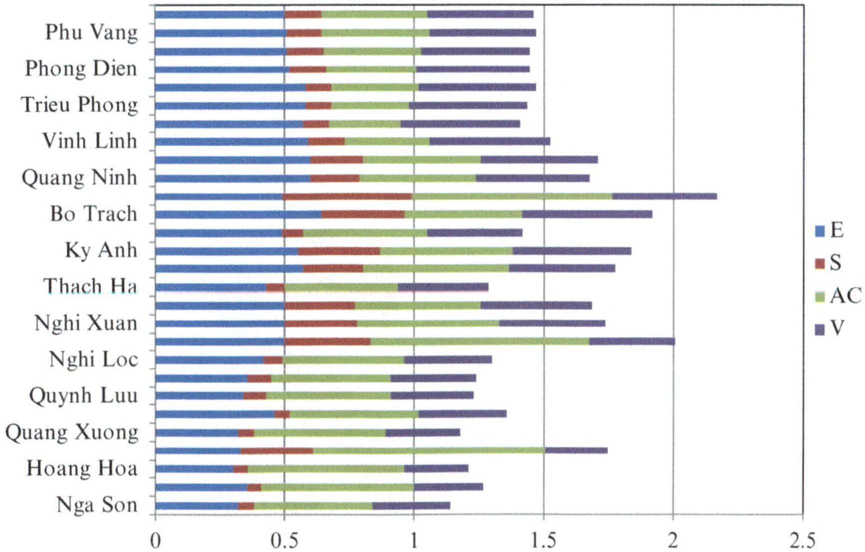

Fig. 2.7 Climate change vulnerability assessment for tourism by district of the Vietnamese Northern Central coast. Horizontal axis indicates value of exposure (E), sensitivity (S), adaptation capacity (AC), and vulnerability (V)

the region is under stress; biodiversity will decrease, and the hot season is expected lasting longer. All this will have a significant impact on the assets and the revenue from tourism. Tourism experiences the lowest vulnerability as compared to the other sectors in the region due to its low exposure. The vulnerability value ranges from 0.2 to 0.5. The Northern Central coast is attractive for both international and domestic visitors. With its diverse natural conditions and special cultures, the region can develop the wide range of tourism types, bathing, sightseeing, and adventure travel, listing just these examples. The tourists can go on tour in the region all over the year, and tourism develops fairly stable. The sensitivity of the sector to the climate variation is low. The highest value of vulnerability was found in the Bo Trach (Quang Binh province), and Sam Son (Thanh Hoa province) scored lowest. In most districts tourism faces moderate susceptibility (27/28), and only two districts in Thanh Hoa province (Sam Son and Hoang Hoa) have a low susceptibility. The Sam Son (Thanh Hoa) shows low vulnerability as result of a high adaptation capacity: the local government invested in the development of the beaches (Fig. 2.7).

In summary, the vulnerability assessment of economic sectors to climate change in the Vietnamese Northern Central coast shows three sectors with a high vulnerability: agriculture, aquaculture, and population. Among them, aquaculture is most vulnerable to climate change. The other sectors have a moderate vulnerability. Tourism shows the least vulnerability. The Quang Tri province suffers most from climate change; in contrast, the Thanh Hoa province is least affected as compared to the other provinces in the region (Fig. 2.8).

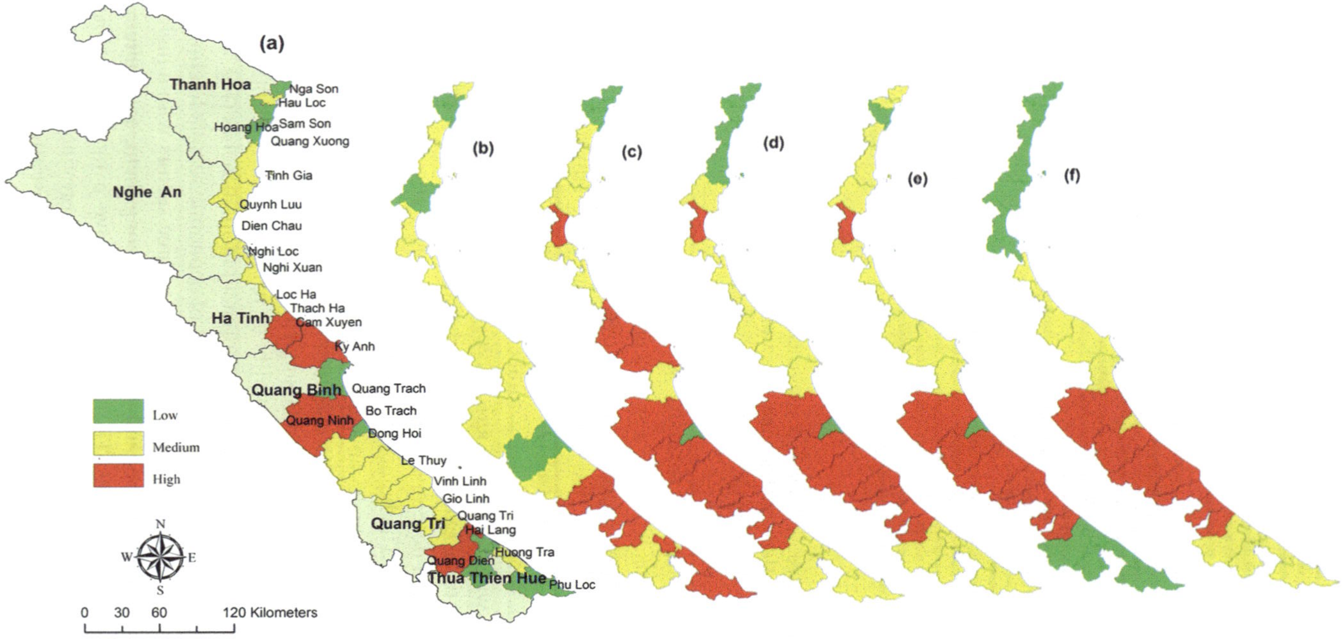

Fig. 2.8 Maps of climate change vulnerability (V) of sectors in coastal districts along the North Central Coast of Vietnam: (**a**) agriculture; (**b**) aquaculture; (**c**) forestry; (**d**) industry; (**e**) population; and (**f**) tourism

2.2 Climate Change Hazards in the Vietnamese Northern Mountainous Region

2.2.1 Vietnamese Northern Mountainous Region

The Vietnamese northern mountainous region (sometime it is called Vietnamese North Mountains and midlands) includes 15 provinces: Ha Giang, Cao Bang, Lao Cai, Bac Kan, Lang Son, Tuyen Quang, Yen Bai, Thai Nguyen, Vinh Phuc, Phu Tho, Bac Giang, Lai Chau, Dien Bien, Son La, and Hoa Binh (Fig. 2.9). Just across the border, three Chinese cities are located. This is the vastest province in Vietnam, covering 100,965 square kilometers or 28.6 percent of the total land surface. Six main landscapes characterize this region: high mountains (Hoang Lien Son mountain range), medium-high mountains (mountain ranges in upstream Ma River), limestone formations (mountain ranges along the Son La, Lai Chau, and Hoa Binh provinces), a mosaic of low mountains and hills (ranges along the Da river valley), valleys (Red River valley), and a mosaic of mountainous landscape (Lo-Gam mountain range). The region witnesses a tropical monsoon climate, with hot and wet summers (April to October) and dry and cold winters (November to March). The average temperature in the region ranges between 18 and 22 °C and 1700 to 2400 mm of precipitation per year. There are 3–6 hot months (average temperature above 25 °C), while the cold lasts for less than 3 months (average temperature below 18 °C) (Le et al. 2012).

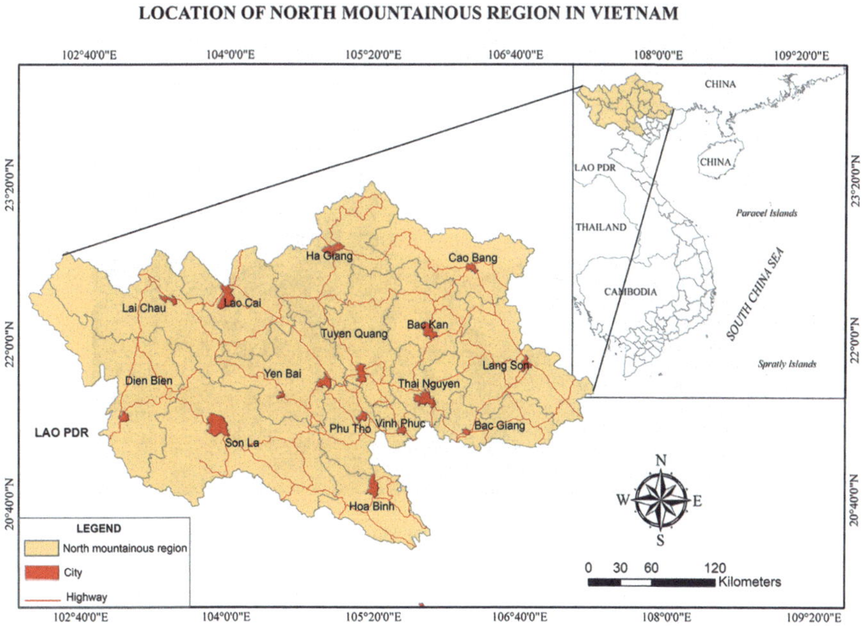

Fig. 2.9 Location of provinces in the Vietnamese northern mountains

The Vietnamese northern mountainous region is sparsely populated (50–100 persons per square kilometers), mostly by Kinh and ethnic minority groups such as Tay, Nung, Hmong, Dzao, Thai, and Muong. Eighty percent of the population's income stems from agriculture, aquaculture, and forestry. The region shows the highest forest cover of the country, 5.6 million hectares, accounting for 42.67% of the national total (GSO 2014). Large plantations of coffee, rubber, pepper, and tea have replaced forests recently due to both locals' activities and the developing policies issued by national, provincial, and local authorities. Agriculture, merely cropping and animal husbandry, provides most of the income. Crop failures and livestock deaths usually occur during the cold weather or excessive rainfall periods. In December and January, the temperature drops below 0 °C; snowfall and hail occur during these cold months. Rain is not evenly distributed throughout the year, as 70% of the precipitation comes in June, July, and August (when the maximum daily rainfall increases to 700 mm). As a result, the region is prone to droughts, cold spells, down pours, landslides, floods, and flash floods, even within the same year (Fig. 2.10).

Fig. 2.10 Thai people's tea fields, in front of the Hoang Lien Son mountain range, Northern Vietnam

2.2.1.1 Climate Change

As shown in Table 2.3, during the period 1958–2014, the Vietnamese northern mountains witnessed temperature increases. Also more abnormal cold events occur. A consistent upward trend in the minimum and maximum temperatures has been described for this area. Most notably, the minimum temperature increased faster than the maximum temperature. The number of cold nights (temperature below the 10th percentile) decreased, while the number of the hot days (temperature above the 90th percentile) increased. During the winter of 2008, a cold period lasted for 38 days in January and February. Temperatures dropped below zero, leading to hail

Table 2.3 Average change in annual temperature (°C) compared to the baseline period 1986–2005 (lower limit of 10% and upper limit of 90%) in the Vietnamese northern mountainous provinces according to two core scenarios

	RCP 4.5			RCP 8.5		
Province	2016–2035	2046–2065	2080–2099	2016–2035	2046–2065	2080–2099
1. Ha Giang	0.6 (0.1–1.1)	1.7 (1.1–2.5)	2.3 (1.5–3.5)	1.1 (0.6–1.6)	2.2 (1.5–3.3)	3.9 (3.1–5.8)
2. Cao Bang	0.6 (0.2–1.1)	1.7 (1.2–2.6)	2.3 (1.6–3.4)	1.1 (0.6–1.6)	2.2 (1.5–3.5)	4.0 (3.1–5.7)
3. Lao Cai	0.7 (0.3–1.1)	1.7 (1.2–2.3)	2.3 (1.5–3.3)	1.1 (0.6–1.6)	2.2 (1.5–3.2)	3.9 (3.1–5.6)
4. Bac Kan	0.6 (0.2–1.1)	1.7 (1.2–2.6)	2.3 (1.6–3.5)	1.1 (0.6–1.6)	2.2 (1.5–3.4)	4.0 (3.1–5.7)
5. Lang Son	0.6 (0.2–1.0)	1.7 (1.2–2.6)	2.3 (1.6–3.3)	1.0 (0.5–1.6)	2.2 (1.4–3.4)	4.0 (3.0–5.6)
6. Tuyen Quang	0.6 (0.1–1.1)	1.7 (1.2–2.5)	2.4 (1.7–3.5)	1.1 (0.5–1.7)	2.3 (1.5–3.4)	4.0 (3.0–5.8)
7. Yen Bai	0.6 (0.2–1.1)	1.7 (1.2–2.3)	2.3 (1.6–3.3)	1.1 (0.6–1.6)	2.2 (1.5–3.2)	3.9 (3.1–5.6)
8. Thai Nguyen	0.6 (0.2–1.1)	1.7 (1.2–2.6)	2.4 (1.7–3.4)	1.1 (0.6–1.7)	2.3 (1.5–3.4)	4.0 (3.0–5.7)
9. Vinh Phuc	0.7 (0.3–1.1)	1.7 (1.2–2.5)	2.4 (1.7–3.5)	1.1 (0.6–1.7)	2.3 (1.4–3.4)	4.0 (3.0–5.7)
10. Phu Tho	0.7 (0.2–1.1)	1.8 (1.2–2.5)	2.4 (1.7–3.5)	1.1 (0.6–1.7)	2.3 (1.4–3.4)	4.0 (3.0–5.8)
11. Bac Giang	0.7 (0.3–1.0)	1.7 (1.2–2.5)	2.3 (1.6–3.3)	1.0 (0.5–1.6)	2.2 (1.4–3.4)	3.9 (3.0–5.5)
12. Lai Chau	0.7 (0.4–1.1)	1.7 (1.2–2.3)	2.3 (1.5–3.3)	1.1 (0.6–1.7)	2.2 (1.4–3.1)	3.9 (3.1–5.5)
13. Dien Bien	0.7 (0.4–1.1)	1.7 (1.2–2.3)	2.3 (1.5–3.3)	1.1 (0.6–1.7)	2.2 (1.4–3.2)	3.9 (3.0–5.6)
14. Son La	0.7 (0.3–1.1)	1.6 (1.2–2.3)	2.3 (1.6–3.2)	1.1 (0.6–1.6)	2.2 (1.5–3.2)	3.9 (3.0–5.6)
15. Hoa Binh	0.7 (0.3–1.1)	1.6 (1.2–2.3)	2.3 (1.6–3.3)	1.0 (0.6–1.5)	2.2 (1.4–3.3)	3.8 (2.9–5.5)

Source: MONRE (2016)

in the northern uplands. Also the 2015–2016 winter brought low temperatures (-5 to -4 °C) and unprecedented snowfall in the uplands and for the first time also in the northern lowlands (MONRE 2016).

The Vietnamese northern mountainous region became drier over the last 50 years. The rainfall tended to decrease by 5.8–7.3% over this period, merely in the fall of the year (by 40%). Rain during spring time tended to increase by 19.5% in the Northwest. Rainfall extremes decreased in most observation stations in the North, while the most pronounced decrease is observed in the North-East (Nguyen et al. 2011). The number of consecutive dry days increased in the North. The maximum rainfall during 1 to 5 days was not consistent in all observation stations; however, there are an increasing trend in the daily maximum rainfall in the North-West and a decreasing trend in the North-East. The consecutive dry days increase all over the country, and the amount of torrential rainfall (95th percentile) decreased at varying rates across the stations.

The relative humidity in Vietnam is high, especially in the mountains (up to 87%). The annual variation of the humidity is correlated with the precipitation pattern. The humidity is low during early and mid-winter, higher during late winter, followed by a period of lower humidity, and more a precipitation during summer (Nguyen et al. 2011).

The mountains in the North of Vietnam will likely face a warmer climate in the future, as temperatures increase during all four seasons. For the mid-century (2046–2064) changes, one expects that on average, the annual temperature in the region will increase by 1.7–2.2 °C, with a range of 1.2–2.5 °C, and 1.4–3.4 °C, in RCP 4.5 and RCP 8.5, respectively. By 2050, the region may experience an increase of the maximum temperature of 2.6–3 °C (RCP 8.5) and 1.4–1.8 °C (RCP 4.5). A similar pattern is expected for the minimum temperature (MONRE 2016).

Table 2.4 shows that annual rainfall tends to increase, ranging from 7.8% (Ha Giang) to 18.8% (Bac Giang). The distribution of the rain is not uniform over the season. Wet deposition decreases in winter time in most provinces by 12.6% (Lai Chau), which points to drier winters. In line with the decreasing winter humidity, winter droughts are expected. Most provinces will experience more rain during summer (a 20% increase in several provinces). As this period coincides with the flood season, also the flood risk will likely increase (MONRE 2016).

2.2.2 *Natural Hazards*

2.2.2.1 Floods, Flash Floods, and Landslides

Floods, flash floods, and landslides occur frequently in Lao Cai, Ha Giang, Lai Chau, Son La, Cao Bang, Bac Kan, and Yen Bai. During the period 2000–2014, about 250 floods, flash floods, and landslides happened in these areas. They caused 646 deaths, destroyed over 9700 houses, and flooded about 100,000 houses. More than 75,000 hectares of rice and other crops were flooded, hundreds of hectares of

Table 2.4 Average change in annual rainfall (%) compared to the baseline period 1986–2005 (lower limit of 20% and upper limit of 80%) in the Vietnamese northern mountainous provinces according to two core scenarios

	RCP 4.5			RCP 8.5		
Province	2016–2035	2046–2065	2080–2099	2016–2035	2046–2065	2080–2099
1. Ha Giang	5.8 (2.7–8.9)	7.8 (3.1–12.6)	11.8 (5.0–19.0)	−3.3 (−9.6–3.3)	4.0 (−0.2–8.1)	12.7 (6.6–18.8)
2. Cao Bang	14.2 (8.2–19.9)	16.0 (9.8–21.8)	22.1 (13.1–31.4)	3.8 (−4.2–11.8)	2.8 (9.4–16.1)	25.7 (17.0–34.4)
3. Lao Cai	1.8 (−4.0–7.1)	8.2 (3.0–13.8)	9.3 (2.2–17.0)	−2.9 (−8.0–2.5)	5.9 (0.4–10.9)	12.6 (5.2–20.0)
4. Bac Kan	17.4 (11.3–23.1)	18.3 (13.5–22.7)	23.7 (16.9–30.8)	6.6 (0.2–13.1)	15.4 (10.4–20.3)	28.0 (19.4–36.1)
5. Lang Son	18.7 (7.0–29.8)	18.7 (11.5–25.5)	25.1 (16.5–34.2)	10.5 (4.6–17.0)	17.9 (12.4–23.3)	27.8 (20.1–35.1)
6. Tuyen Quang	11.5 (6.2–16.4)	12.5 (7.5–17.7)	18.4 (10.2–27.1)	5.8 (−0.1–11.6)	16.7 (9.7–23.5)	27.4 (15.0–38.7)
7. Yen Bai	7.5 (0.2–14.3)	14.8 (7.5–23.0)	19.4 (7.8–32.7)	5.9 (−0.7–12.7)	15.6 (7.9–23.3)	23.3 (9.4–35.7)
8. Thai Nguyen	15.9 (8.2–23.3)	17.8 (11.1–24.2)	22.5 (14.9–31.0)	9.9 (4.9–15.0)	22.0 (13.8–30.2)	31.1 (21.8–40.1)
9. Vinh Phuc	14.8 (5.4–24.6)	18.2 (10.6–26.6)	22.4 (12.5–34.1)	10.7 (4.7–17.0)	22.2 (12.4–32.1)	30.8 (18.5–42.1)
10. Phu Tho	10.0 (0.3–19.7)	15.0 (8.2–22.6)	21.3 (10.7–33.4)	8.5 (1.6–15.6)	17.1 (7.5–26.1)	25.4 (11.8–37.4)
11. Bac Giang	17.7 (5.4–29.3)	18.8 (11.0–26.9)	25.7 (16.6–35.6)	10.9 (5.8–16.7)	21.1 (15.4–27.2)	32.7 (25.5–39.5)
12. Lai Chau	3.3 (−3.3–9.7)	13.5 (9.4–17.9)	11.2 (4.6–18.3)	−1.0 (−4.0–2.1)	10.6 (4.4–16.0)	18.4 (12.0–25.3)
13. Dien Bien	5.9 (−2.2–13.2)	16.5 (8.9–24.3)	15.1 (6.6–24.4)	2.7 (−1.7–7.3)	15.2 (8.0–21.7)	21.2 (14.8–28.2)
14. Son La	7.0 (−0.5–14.2)	7.0 (−0.5–14.2)	19.9 (10.3–30.4)	5.1 (−1.3–11.2)	15.3 (9.3–21.3)	22.3 (15.7–28.9)
15. Hoa Binh	7.5 (0.0–15.4)	12.9 (8.1–18.1)	20.2 (12.2–29.1)	7.0 (1.4–12.9)	12.8 (7.4–18.2)	20.9 (12.4–29.0)

Source: MONRE (2016)

farmland were covered, and transportation and irrigation infrastructure was impaired. The total damage was estimated at 150 million $US (CCFSC 2014). A list of major floods, flash floods, and landslides over the years 2000–2017 is presented in Table 2.5.

The damage caused by flash floods and landslides in the Vietnamese northern mountains is mainly related to heavy rainfall during a short time on steep slopes. Cases were reported where the damage is the result of the economic development with limited prevention and risk management:

Table 2.5 Damage of major floods, flash floods and landslides in the Vietnamese northern mountainous regions since 2000

Natural hazard	Period	Place	Damage
Floods	July 2000	Sa Pa district (Lao Cai province)	20 deaths
Flash flood	October 2000	Nam Coong village (Nam Cuoi commune, Sin Ho district, Lai Chau province)	39 deaths
Flash flood	August 2002	Bac Quang and Xin Man districts (Ha Giang province)	25 deaths
Floods	2004	Du Tien and Du Gia communes (yen minh district, Ha Giang province)	45 deaths
Landslides	September 2004	Sung Hoan village (Phin Ngan commune, bat Xat district, Lao Cai province)	20 deaths
Flash floods	September 2005	Cat Thinh commune (Van Chan district, Yen Bai province)	22 deaths
Floods, flash floods, landslides	2008	Lao Cai and Yen Bai province	120 deaths
Landslides	2009	Pac Nam commune (Bac Kan province)	13 deaths
Landslides	2010	Hang Tau Do village (Che Cu Nha commune, Mu Cang Chai district, Yen Bai province)	7 deaths
Flash flood	September 2013	Ban Khoang commune (Sa Pa district, Lao Cai province)	11 deaths
Floods and landslides	2014	Ha Giang, Lai Chau, Cao Bang, Son La provinces	24 deaths
Floods, flash floods, and landslides	August 2017	Van Chan and Mu Chang Chai districts (Yen Bai province), Muong La district (Son La province)	12 deaths

Source: Vietnamnet, accessed on 15 August 2017 at http://vietnamnet.vn

- Unsustainable road construction resulting in slides.
- Over-exploitation of natural resources, construction of bridges, irrigation, and flood drainage.
- River filling and leveling for construction of infrastructure and houses and production facilities caused flow congestion and narrow streams.
- Location of new houses along streams, on slopes, foothills, and mountains, close to traffic roads, or in low-lying areas in valleys.
- Limited awareness among the locals on the prevention of natural disasters. For example, falls while crossing river and lack of attention during floods (Fig. 2.11).

2.2.2.2 Extreme Colds and Drought

Extreme colds impact the lives and livelihood of people in the Vietnamese mountains. Rice, crops, and livestock are vulnerable. Cold temperature during long period occurred in 2008 and 2010. Thousands of cattle and buffaloes died, while thousands hectares of rice and crops were lost. In the Bac Kan province, the extreme cold in

Fig. 2.11 A combination of floods, flash floods, and landslides occurred in the Van Chan mountainous district (Yen Bai province, Vietnam) on August 2017

2008 destroyed about 100,000 hectares of rice and killed livestock, causing a loss of about 2.2 million $US (BKG 2009). In the Lao Cai province, over 84 tons of rice was lost as a result of the extreme cold in 2008. Cold periods not only affect the rice production but also forestry, fruits, and vegetables. In general, extreme colds have a direct impact on food crops, cash crops, fruit trees, and forestry yield and result in the loss of arable land (LCG 2009). For the Ha Giang province, the total loss caused by natural disasters in 2010 was estimated at 6.6 million $US, of which extreme colds caused nearly 4.4 million $US, mainly as a result of lost livestock and rubber trees (HGG 2011).

In 2011, long droughts mainly damaged agriculture. Only 93% of the cultivation land foreseen by the provincial agricultural plan was used. 1820 hectares of rice paddies were affected by drought, of which, 120 hectares of rice were replanted and 20 hectares of rice paddies were totally lost (CCFSC 2014).

2.3 Conclusion

Overall both larger areas in which the case studies are localized experience increasing temperatures. Along the coast the mean air temperature during summertime faces today an increase of 0.7 °C as compared with the mean during the last decades of the previous century. This figure is expected to increase to 1.9–3.3 °C by the end of this century. Also for the mountains, a consistent upward trend of the air temperature is forecasted.

Along the coast one expects more rain. The gentlest scenario expects for the next two decades a 10–17% increase in precipitation as compared to the last decades of the last century. This figure might (almost) double by the end of the century. In the mountains the forecasted precipitation increase of 0.7 °C might triple by the end of 2100. Remarkably this increase is expected to be accompanied by more and longer dry periods.

These rough indicators of a changing climate are symptomatic for the effects they cause. Although the number of storms originating over the sea decreased during recent years, the storm period extends, and the storms are heavier, causing increasingly more damage. In the mountains floods, flash floods, and landslides are the main impacts, both in terms of human deaths and damage to infrastructure. In the coastal territories, agriculture, aquaculture, and people are most vulnerable, while the tourism sector shows the least vulnerability. These regional descriptive-analytical data and the consequences of the meteorology findings provide the necessary context for the impacts on people and nature which are the subject of the chapters in part 2 of this book.

Part II
Risks for Nature and Humans in the Lowland

Chapter 3
Climate Change Hazards and Migration Along the Ky Anh Coast

Abstract This chapter narrows the geographical scope of the previous one and addresses in more detail the effects of natural hazards on the coast of one of the most affected areas by storms in Vietnam the coast of Ky Anh. It focusses on the coastal communes in the district which border the South China Sea. The introductory sections provide a general description of the communes including an analysis of the land use changes during the last two decades. The area faces tropical storms and is affected by beach erosion. A review of the main recent storms and floods and the hazards they bring for the locals living in the dunes near the sea is provided. Coastal erosion is measured along transects in the selected communes during four decades. Total gross migration figures are established based on the information by the local authorities, for each of the communes. Their likely relation with the storms, floods, and flash floods is analyzed.

Keywords Climate change hazard · Tropical storm · Coastal erosion · Migration · Land use land cover · Ky Anh coast

3.1 The Coast of Ky Anh

Ky Anh coast is part of Ha Tinh, which is the most northern province of Central Vietnam (Fig. 3.1). The Ha Tinh province is located east of the Truong Son mountain range, which runs from north to south and forms an almost physical border with Laos PDR in the west. The South China Sea is east of Ha Tinh. Just like other provinces in Central Vietnam, the coast of Ha Tinh is characterized by merely wide and sandy beaches bordered by a forested dune ridge (Fig. 3.2). The lowland behind the dunes is for the farmers and their fields. This flat landscape is dotted with sandstone hills, which do not exceed 800 meters. Also, the remaining part of the province is hilly with mountains below 1000 meters above sea level (Le et al. 2012). Since its reestablishment in 1991, the Ha Tinh province developed rapidly. Particularly during the most recent 10 years, urbanization and industrialization lead to expanding urban areas (cities and county towns), industrial zones, and transport facilities. The Ha Tinh province had 1.2 million inhabitants in 2012. Ha Tinh city is the capital of

A. T. Nguyen, L. Hens, *Human Ecology of Climate Change Hazards in Vietnam*, Springer Climate, https://doi.org/10.1007/978-3-319-94917-8_3

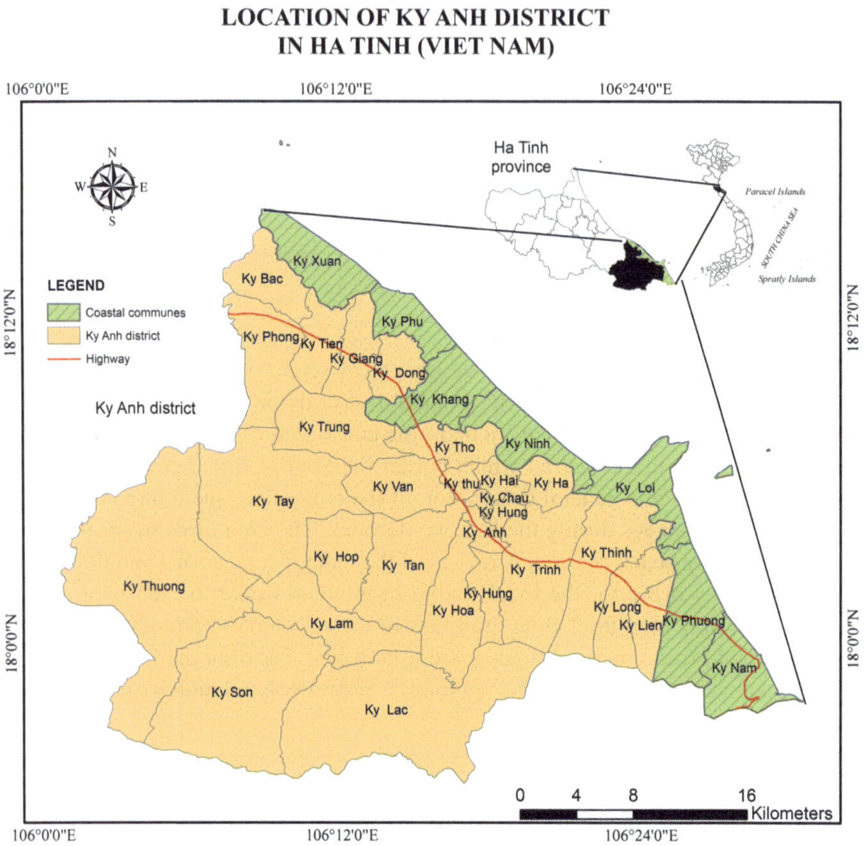

Fig. 3.1 Location of the Ky Anh coast in Ha Tinh province (Vietnam)

the province and hosts about 62,000 inhabitants; most of the rural inhabitants are farmers who live mainly along the coast and in the hills (HTG 2013).

The likely effects of climate changes are most tangible in this province. They include:

1. The average temperature during the period 2000–2010 increased by 0.6 °C as compared to the period 1970–1980.
2. Extreme weather events: Unusual cold periods (the spring of 2009 was the coldest of the last 40 years) alternate with heat waves (in July 2010, the province experienced during 10 consecutive days temperatures over 40 °C); storms are frequently accompanied by heavy rains (the 2010 flood lasted for more than 20 days).
3. Changes in the frequency, the timing, and the intensity of the tropical storms are part of the changing weather profile. While traditionally storms occurred during the period September–November, the storm season now extends from August to December. Floods occur from April to December. They become stronger and faster, with more peak events and more devastating impacts (IPONRE 2009).

Fig. 3.2 Coastal landscape of Ky Anh. (Source: An Thinh Nguyen 2017)

In short, prolonged periods of high and low temperatures, drought, sea level rise, storms heavy rains, and (sudden) floods are considered the main weather drivers affecting the livelihood of these communities in coastal Ha Tinh. Consequently, Ha Tinh faces four main problems:

1. *Changes in water supply:* Drinking water supply and irrigation are critical all over the province. In 2010, 27% of the agricultural land was irrigated. The provincial policy goal is irrigating 70% of the fields. Also by 2010, 70% of the population had access to piped water. The daily per capita consumption ranges from 80 to 100 liters on average. The policy goal is supplying 100% of the urban and 80 to 90% of the rural population with safe drinking water (HTG 2013). The increasing pressure on the water supply hampers realizing these goals.
2. *Changing land use and urbanization:* By 2001, 10% of the land in Ha Tinh was urban area, while the remaining surface was rural. By 2010, the urban land covered 15% of the province, while the rural area decreased to 85% (HTG 2013). The figures illustrate the conversion of agricultural and bare land into urban areas. Consequently, the area is also increasingly affected by the urban heat island effect.
3. *Progressing shoreline erosion:* Depending on the inclination of the beaches, Ha Tinh loses beaches at a rate of 0.2–15.0 meters per year.
4. *Changing livelihoods:* Both urbanization and the changing climate affect the way of life in Ha Tinh. Especially farmers, aquaculturists, and fishermen change

their habits, adapting to the increasing storms. Urbanization is associated with changes in consumption lifestyles, the size of the families, the ways of commuting, the gender roles, and the time residents spent at home.

Ky Anh is the utmost southwestern district of the Ha Tinh province as shown in Fig. 3.1. The district covers 105,428 hectares and entails 7 coastal communes from Ky Xuan in the north to Ky Nam in the south. These communes span 63 kilometers of coastline, with mainly sandy and to a lesser extent rocky, and beaches. Behind the dune ridge, the delta is covered with irrigated rice fields. A more limited, dryer area allows for growing peanuts and vegetables. The fields are dotted with low hills, which are covered by conifers, mixed leafy trees, and scrub vegetation. The rivers are bordered by (merely planted) mangroves. Ky Anh has lakes, one of which is artificial, providing water for the irrigation of rice.

By 2012, the Ky Anh district totaled 174,216 residents (51,356 households), 48,025 of whom were farmers and fishermen (HTG 2013). While the population and the number of households increased (the total population increased from 171,784 in 2005 to 174,216 in 2012, and the number of households raised from 41,806 in 2001 to 51,356 in 2012), the number of farmers and fishermen declined significantly (from 60,445 in 2005 to in 48,025 in 2012). This indicates a noticeable change in the occupation of the residents in the district.

The selected coastal zone of the Ky Anh district covers 105,428 hectares, spreading over 7 coastal communes from north to south along the South China Sea: Ky Xuan, Ky Khang, Ky Phu, Ky Ninh, Ky Loi, Ky Phuong, and Ky Nam. Agroforestry and fishery are the main economic sectors in Ky Xuan, Ky Khang, and Ky Phu, whereas industry is the main economic sector in the Ky Loi and Ky Phuong.

- The Ky Xuan commune has a coastline of 15 kilometers. It was the home of 6512 residents (1882 households) in 2012. Its economic sectors include agriculture (317 ha of mainly rice paddies), marine fish catch, and titan and marble mining. 1312 hectares is planted forest (KAG 2012). Out-migration (the population declined since 2006, when 6833 people leaved Ky Xuan) of mainly young people looking for jobs in the urban areas of Vinh City and Hanoi is a persistent problem. A second reason for the out-migration is that people are displaced from regularly flooded areas (KAG 2009, 2014).
- Ky Khang is a commune south of Ky Xuan, with 5 kilometers of coast. Since 2012, it has a fairly stable population of 11,499 residents (2966 households). The agricultural land of 728 hectares makes it the largest rice producer of the district. 1055 hectares are covered by planted forest. The local economy relies heavily on rice and peanuts, titan mining, and forest products. The local sea fishing company owns 120 vessels (KAG 2012).
- The Ky Phu commune is the smallest one, covering 1528 hectares. People live behind the dune ridge of the 7 kilometers long sandy beaches, which is the longest among the coastal communes of Ky Anh. The number of residents gradually declined from 10,040 people in 2006 to 9688 people (2568 households) in 2012. There are 176 hectares of mainly planted forest in Ky Phu. Most of the villagers (2053/2568 households, or 79.9%) are farmers, cultivating 598 hectares

of fields (KAG 2012). Similar to the other communes, the main crop in Ky Phu is rice and to a lesser extent peanuts, vegetables, and fruits. Shrimp farming, shells, and freshwater fish aquaculture are localized along the rivers and the seashore.

- The Ky Ninh commune extends over 2120 hectares; agriculture and aquaculture cover 954 hectares (KAG 2012). The terrain of the commune is relatively complex, running from northwest to the southeast.
- The Ky Loi and Ky Phuong communes belong to the Vung Ang EZ, which was established in 2006 and is one of the eight special economic zones in Vietnam (VNG 2007). It covers 2056 hectares in Ky Loi and 33,542 hectares in Ky Phuong (KAG 2012).
- In Ky Nam, a new urban center is planned. The commune is mainly bordered by mountains and has 54 hectares of aquaculture and irrigated fields (KAG 2012).

The Ky Anh coastal area was selected for the national key point industrial area. The Vung Ang EZ offers port services for industry, business, tourism, agriculture, and aquaculture. Due to it extending over 4 out of the 7 coastal communes (Ky Ninh, Ky Loi, Ky Phuong, and Ky Nam), an estimated 1100 households have been relocated, and the entire Ky Phuong commune was reclaimed for the construction of the economic zone up to 2014 (VNG 2007).

Along the Ky Anh coast, a variety of mineral mining sites include titan, brown coal, and sand in Ky Xuan, Ky Phu, Ky Khang, and Ky Nam communes. The titan mine has an estimated reserve of 2.1 million tons of ore, 1.1 million tons of which has been exploited by the Ha Tinh Minerals and Trading Corporation (MITRACO) and several other small enterprises. The total production of mainly traditional titan in Ky Anh is 100,000 tons per year (KAG 2012). Most of the titan ore is found under sand banks and dunes along the coast, close to the water edge. Titan mining seriously affects the coastline and beaches. The construction sand extraction reduced the alluvium and exacerbated coastal erosion of both the mining and the neighboring area. To a limited extent, mine reclamation was done, such as in the Cao Thang hamlet of the Ky Xuan commune. As the soil used for reclamation is often loose with much less minerals and consistency, it easily eroded in the rainy season. Titan mining often adversely affects the protected forest (e.g., Casuarina forests stabilize the sand and prevent it from inland transport) (Fig. 3.3).

Illegal sand mining is widely spread common in Ky Anh. Construction sand is merely found in rivers and along the sea. Yearly about 30,000–35,000 m^3 of the river sand area mined in Ky Anh (KAG 2012). The Rao Tro River, which flows through the other inland communes of the Ky Anh district (Ky Lam, Ky Son, Ky Thuong, and Ky Lac), has been threatened for many years by illegal sand mining. This causes damage to the riverbed, changes the flow of the river, and speeds up the erosion of the riverbanks. Illegal sea sand mining is increasingly drawing attention along the coast. These activities significantly reduce the alluvial deposits and increase the seawater intrusion. The topographical impacts along the coast of Ky Anh concern sand mining for sea encroachment and the seaport construction projects in Son Duong. This causes severe sediment loss of the seabed. The sediment flows change both inland and offshore, which destroys the balance of the seabed topography.

Fig. 3.3 The reclaimed titan mine in the protected forests along the Ky Anh coast in 2014. (Source: An Thinh Nguyen 2014)

The change detection analysis shows significant changes of LULC in the study area in two different periods (from 1993 to 2003 and from 2003 to 2013). The built-up land increased over time (with 13,980 hectares during the period 1993–2013). This phenomenon is most pronounced between 2003 and 2013. This increase also applies to the industrial area: new industries have been identified in the classification of the 2013 LANDSAT image, for example, in the Vung Ang EZ. Due to the conversion to built-up land, agricultural land witnessed a steep decrease of paddy fields over time (approximately 13,840 hectares), in particular during period 2003–2013 (nearly 8884 hectares). Forest decreased in the period 1993–2003 (over 2301 hectares); however, it increased since 2003 mainly due to afforestation (over 557 hectares). Table 3.1 and Figs. 3.4, 3.5, and 3.6 show a more complete picture of the LULC change of the study area.

3.2 Tropical Storms and Coastal Erosion

3.2.1 Hazards from Tropical Storms

During the past decades, the coast of Ky Anh suffered from a severe climate with tropical storms, dry hot western wind, drought, heavy rain, and flash floods. All these natural hazards associated with global climate change increased both in

Table 3.1 LULC during 1993–2013 in the Ky Anh coast

LULC \ Year	1993	2003	2013	1993-2003	2003-2013
Built-up	11049.03	17093.41	25028.91	6044.38	7935.5
Agriculture	40290.48	35334.99	26451.09	-4955.49	-8883.9
Forest	42684.48	40383.11	40940.46	-2301.37	557.35
Water	112628.43	113840.91	114231.96	1212.48	391.05
Total	206652.42	206652.42	206652.42		

Source: LULC detection from LANDSAT images; see Box 1.2, Chap. 1

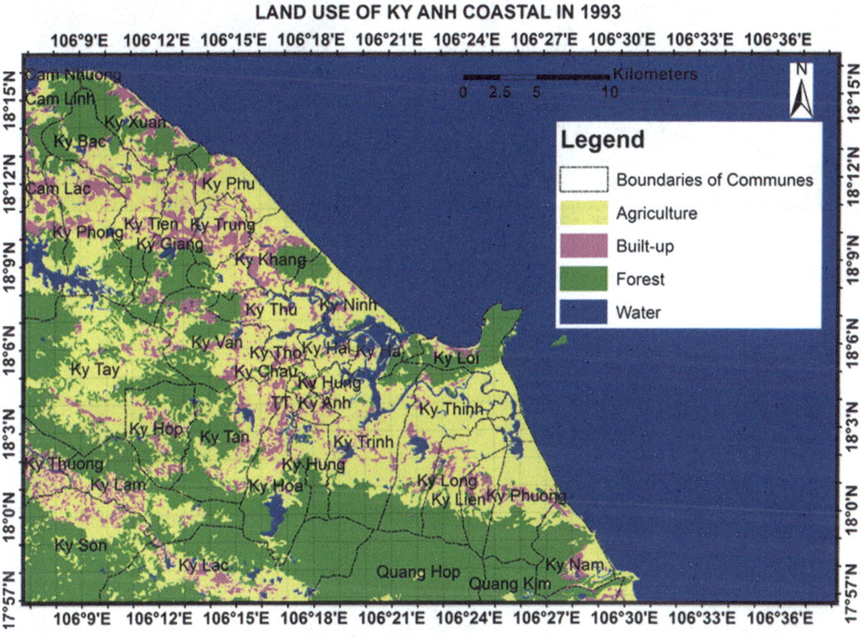

Fig. 3.4 LULC map of the Ky Anh coast in 1993

frequency and in strength (IPONRE 2009). The area is frequently affected by tropical storms in a differential way. Their frequency increases from January to August, to decrease afterward reaching their lowest incidence in December. During the past 50 years, the Ha Tinh province was affected by 47 storms, 18 of which directly hit the coast of the Ky Anh. On average Ha Tinh copes with 0.9 storms per year (IPONRE 2009). Storms cause floods in the Ky Anh coast. The intensity of the floods and the damage they cause worsen by sea level rise. The storm Wutip of September 2013 was the most devastating during recent years. The Mekkhala storm of 30 September 2008 and the Ketsana storm of 29 September 2009 brought heavy rains but limited floods. The Lionrock storm of 27 August 2010 was accompanied by heavy rains. The Nockten storm of 30 July 2011 and the Haitang storm of

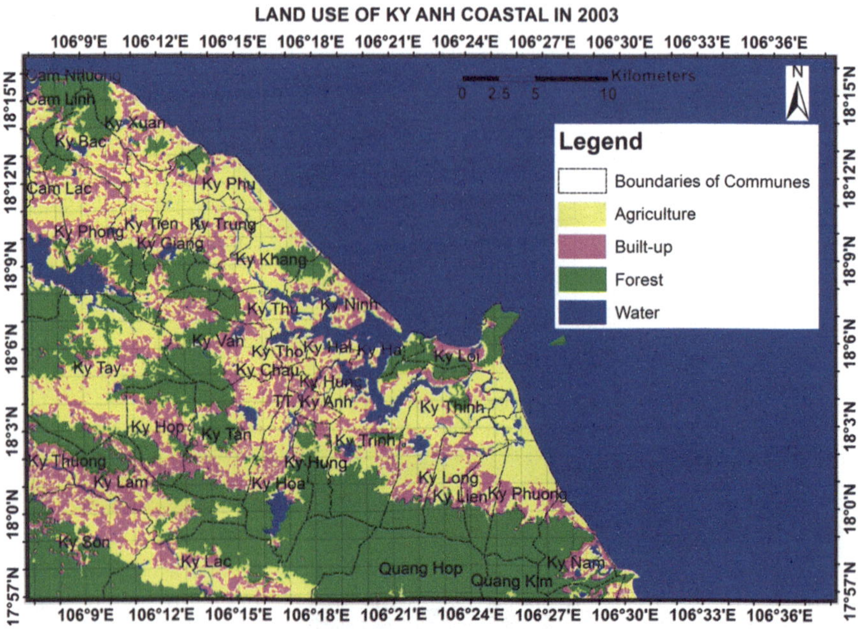

Fig. 3.5 LULC map of the Ky Anh coast in 2003

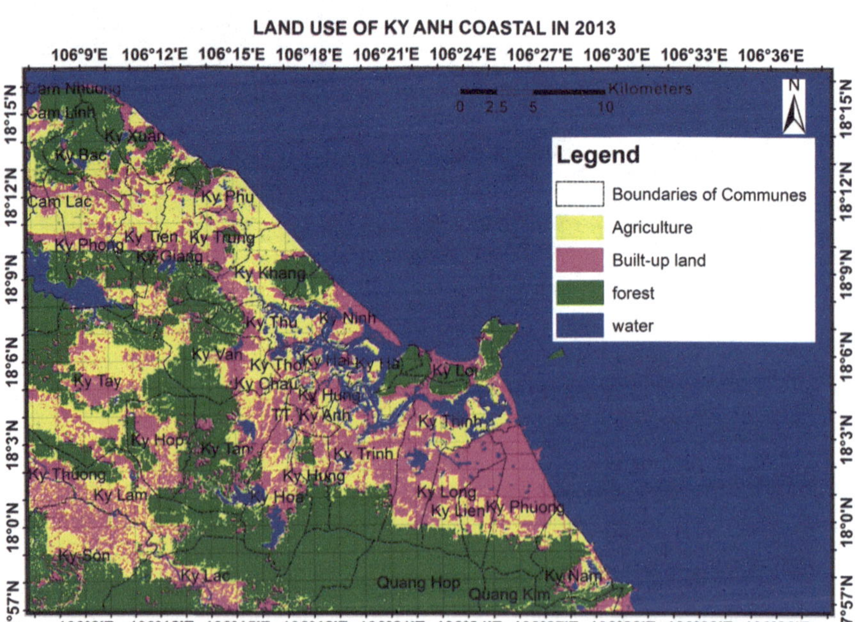

Fig. 3.6 LULC map of the Ky Anh coast in 2013

Table 3.2 Major storms which impacted directly the Ky Anh coast during 2008–2013

No	Year	Storms	Impacts
1.	2013	Wutip (25 Sept 2013) Haiyan (4 Nov 2013)	Severe floods following the storm
2.	2012	Son-Tinh (28 Oct 2012)	Heavy rain with limited floods
3.	2011	Nockten (30 Jul 2011) Haitang (27 Sep 2011)	Heavy rain during the storm
4.	2010	Lionrock (27 Aug 2010) Megi (16 Oct 2010)	Heavy rain during the storm
5.	2009	Ketsana (29 Sep 2009)	Heavy rain with limited floods
6.	2008	Mekkhala (30 Sep 2008)	Heavy rain with limited floods

Source: National Center for Hydro-Meteorological Forecasting (2014); http://www.thoitietnguy-hiem.net

27 September 2011 brought heavy rains but no floods. In contrast the Wutip storm of 25 September 2013 and the Haiyan storm of 4 November 2013 caused major floods (Table 3.2).

The storm winds and the rains not only cause damage but also intensify related hazards as sea level rise, floods, and coastal erosion. The wind speed of the storms ranges from 16 to 30 meters per second, whereas maxima reach 40 meters per second once the storm directly hits the study area (e.g., in 1964). On average, heavy rains (150–250 mm per day) usually accompanied by cold air from the north, characterize the storm weather (IPONRE 2009). Before 2000, coastal erosion due to high waves progressed by about 0.2–1 meter per year (in the Ky Xuan commune). During 2008–2013, the phenomenon increased to about 4–5 meters per year (in the Ky Khang commune), 3–4 meters per year in the Ky Xuan, and 10–20 meters per year in Ky Phu. Drought is less characterized by dramatically increasing temperatures, significant decreasing amounts of rainfall, or long periods of water shortage; it is rather associated with variations in the onset and the end of the wet season. Consequently the effects are mitigated, and the resilience is increased by the construction of dams, protecting the water reserves for the irrigated rice paddies, and by moving to drought resistant crops as peanuts. Nevertheless, both aqua- and agricultures face increasing periods of water shortage.

Floods occur regularly. Usually the flood season starts in August, and it is most frequent in September, October, and November. Flooding levels reach 1.5 meter height on average; they affect 5–25% of the population in the study area. The floods caused by the September 2013 storm submerged parts of the communes with 0.8–2.5 meters of water. Prior to the floods, people were relocated, away from the areas at risk. In 2013, 5–25% of the families had to move temporarily. Prior to the September 2013 storm, 302 households in three coastal communes (Ky Xuan, Ky Khang, and Ky Phu) were evacuated.

Overall, the core data on floods, storms, and sea level rise along three communes of the Ky Anh coast are summarized in Table 3.3.

Table 3.3 Characteristics of hazards from tropical storms in three coastal communes of the Ky Anh district during 2008–2013

Hazards from tropical storms	Ky Xuan	Ky Khang	Ky Phu
1. *Flood*			
1.1. Season/period	Apr to Dec	Apr to Dec	Apr to Dec
1.2. Most recent (data)	Sept 2013	Sept 2013	Sept 2013
1.3. Maximum height (meters)	0.8	1.6	2.5
1.4. Paddy area flooded (% of total area)	15–20	70	50
1.5. Citizens temporarily relocated (number of households)	98	113	123
2. *Storm*			
2.1. Frequency (per year)	0.9	0.9	0.9
2.2. Most recent main storm	Sept 2013	Sept 2013	Sept 2013
2.3. Most affected households	301	139	450
2.4. Citizens relocated	17	130	155
2.5. Number of the most affected villages	4	2	2
3. *Sea level rise*			
3.1. Rise during the most recent decade (2003–2013) (cm)	20–25	20–25	20–25

3.2.2 Coastal Erosion in the Period 1973–2013

Coastlines of Ky Anh area in 1973, 1987, 1994, 2003, and 2013 and the corrected tidal level changes demonstrate complex variations of different shapes, extents, and trends. The corrected shorelines were extracted from LANDSAT images. The coast of the seven communes was divided into seven sectors. A transect was defined to determine the value of the change during the observation period. Transects AA, CC, FF, and GG are drawn for the Ky Xuan, Ky Khang, Ky Loi, and Ky Nam communes, where erosion is manifest at the observation date (Fig. 3.7b, d, g, and l). Transects BB, DD, and EE were set up for the Ky Phu and the Ky Ninh communes, where sand deposition was found (Fig. 3.7c, e, and f). Transect EE, a shoreline north of the Tri river mouth, eroded. The shoreline of the Son Duong port and the integrated steel mill complex in the Ky Phuong commune suddenly changed between 2010 and 2013 (Figure 3.7h).

To analyze the shoreline changes during 40 years, seven transects are defined and plotted (Fig. 3.8). The seven transects are represented by seven colors with their error bars (± 1 pixel of LANDSAT MSS, TM, and OLI data). The shoreline of 1973 provided the baseline, while the others are compared with the baseline to measure the difference. If positive, the value of change coincides with deposition, whereas a negative value points to erosion. The results in Fig. 3.8 show overall erosion. More in detail, they correspond with the following shorelines changes:

- *Ky Xuan commune:* Transect AA shows erosion during the periods 1973–1987 and 2003–2013. The shoreline moved about 120 meters. Erosion was the main

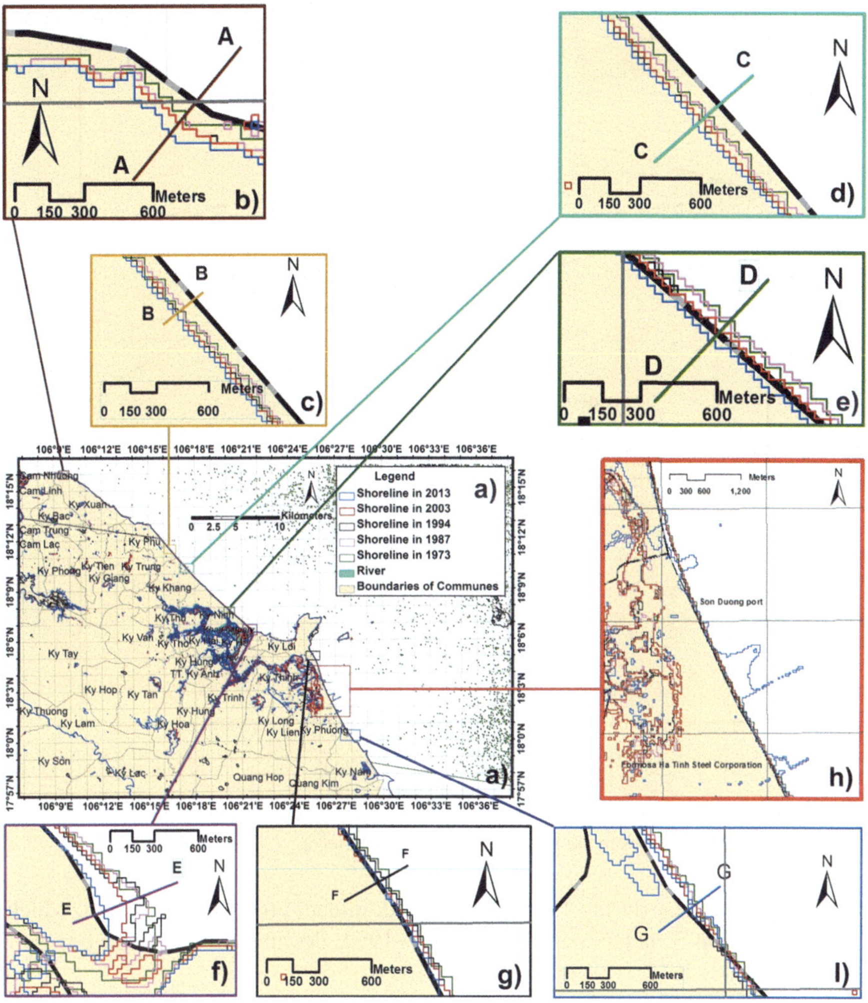

Fig. 3.7 (a) Shorelines of the Ky Anh coast along five transects; (b) transect AA in Ky Xuan; (c) transect BB in Ky Phu; (d) transect CC in Ky Khang; (e) transect DD in Ky Ninh; (f) transect EE in the Khau estuary; (g) transect FF in Ky Loi; (h) Son Duong port and steel factory in the Vung Ang EZ; and (I) transect GG in Ky Phuong

trend, with an average rate of approximately 3 meter per year. The average is the result of 12 meter erosion per year during 2003–2013, while there was no change, in the years 1987–2003.

- *Ky Phu commune:* The shoreline changed from accretion during 1973–1987 to erosion during the period 1987–2013, as shown by transect BB. The shoreline moved about 37 meters in 1973–1987; this was less during 1987–1994 and increased to −20 meters during 2003–2013. The highest accretion rate was about

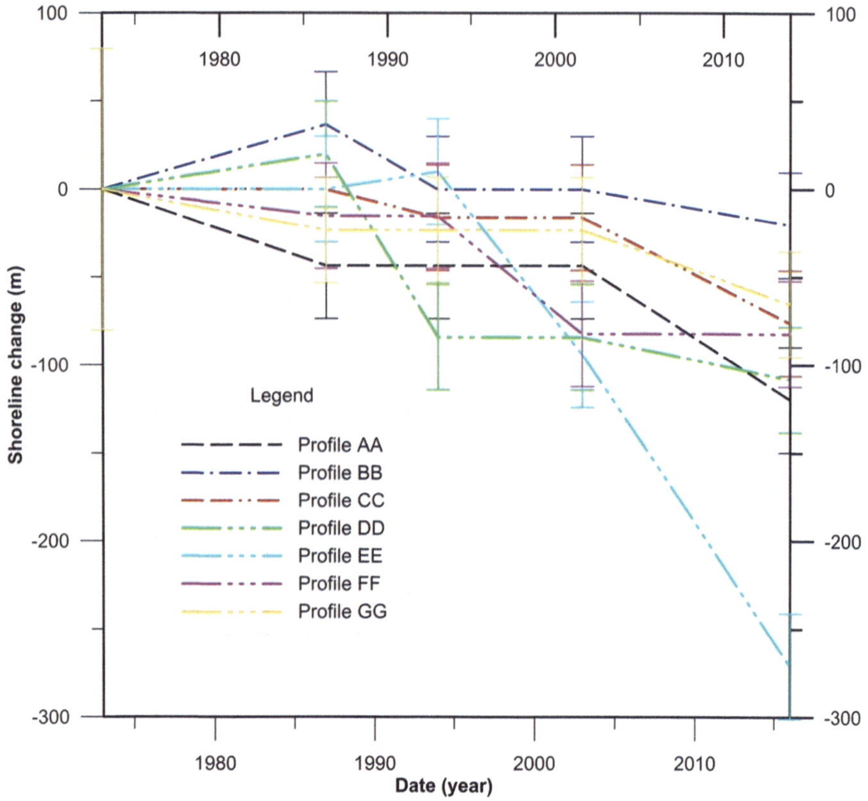

Fig. 3.8 Shoreline changes in the Ky Anh coast along transects

3 meters per year during 1973–1987. The dominant process was erosion, which progressed at 3 meters per year in 1987–1994, declining to 2 meters per year during 2003–2013.

- *Ky Khang commune:* The shoreline shown in transect CC shows changes of −16 meters in 1987–1994, to −76 meter in 2003–2013. Erosion of 6 meters per year occurred during 2003–2013. However, no change was observed in 1973–1987 and 1994–2003.
- *Ky Ninh commune:* The shoreline slightly accredited during 1973–1987 but eroded during the period 1987–2013 as shown in transect DD. The shoreline change ranged from 20 meters in 1973–1987 but decreased to −104 meters in 1987–1994 and − 24 meters during 2003–2013. The highest accretion rate was 1 meter per year during 1973–1987. The dominant process was erosion, varying from 15 meters per year in 1987–1994 to 3 meters per year during 2003–2013.
- *Ky Loi commune:* Transect FF in 1973–1987 and 1994–2003 showed changes of about −15 meters and −67 meters, respectively. The fastest erosion rate was approximately 7 meters per year in 1994–2003. During 1987–1994 and 2003–2013, the shoreline remained stable.

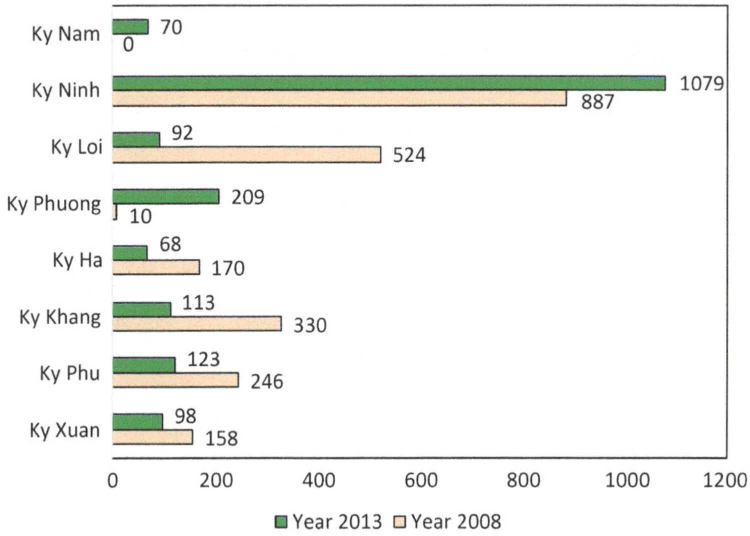

Fig. 3.9 Houses were damaged by a combination of a tropical storm, which also caused flood and sea level rise along the Ky Anh coast in 2013. All households here were relocated. (Source: Nguyen An Thinh 2013)

- *Ky Phuong commune:* Accretion occurred as the result of the construction of the Son Duong port and the steel factory in the Vung Ang EZ. Here 4100 square meters of land surface increased along the shoreline. The largest increase was about 823 meters in 2003–2013.
- *Ky Nam commune:* The coastal zone in Ky Nam shows less erosion than other coastal areas during 1973–2013 as illustrated in transect GG. The most dramatic change of −42 meters occurred during 2003–2013; a decline by −23 meters was found during the period 1973–1987; during 1987–2003, the coastline remained stable. The highest erosion rate was 4 meters per year. The dominant process was slow erosion.

3.3 Climate Change, Migration, and Adaptation

Along the coast of Ky Anh, climate change hazards impacted most seriously the settlements and relocation of locals. In 2008, the Ky Nam commune is the only one which is rather unaffected by these natural hazards and consequently by migration (KAG 2009). The six remaining coastal communes are affected by the storm-induced hazards, which resulted in migration. Local people living in these communes change their home before, during, and after the floods (Fig. 3.9).

Migration was most pronounced in Ky Ninh as most of its territory is low-lying and hardly drains after long periods of rain. Here the impacts of floods appeared

Table 3.4 Out-migration from natural hazard damaged areas of the Ky Anh costal area in 2008

Coastal communes	Flash floods			Flooding		
	Number of affected households	Number of affected people	Rate (%)	Number of affected households	Number of affected people	Rate (%)
Ky Xuan	158	634	12.22	0	0	0.00
Ky Phu	246	993	13.18	0	0	0.00
Ky Khang	330	1020	15.09	0	0	0.00
Ky Phuong	0	0	0.00	10	40	1.05
Ky Loi	0	0	0.00	524	1907	32.87
Ky Ninh	0	0	0.00	887	2025	71.24
Ky Nam	0	0	0.00	0	0	0.00

Source: KAG (2009)

Table 3.5 Migration in hazard-impacted areas along the of coast of Ky Anh in 2013

Coastal communes	Type of hazards (number of migrant households)			
	Flash floods and floods	8–9 force gales	10–11 force storms	12 force hurricanes
Ky Xuan	98	17	56	301
Ky Phu	123	155	257	450
Ky Khang	113	130	130	139
Ky Phuong	209	0	0	86
Ky Loi	92	70	456	900
Ky Ninh	1079	379	715	875
Ky Nam	70		30	50

Source: KAG (2014)

faster and were more serious than those of flash floods. Flash floods most frequently happen in the mountains and the hills, while water levels increase all over the region. The migration rate in the Ky Phuong and Ky Nam is limited: 1.05% and about 0%, respectively (Table 3.4).

As shown in Table 3.5, households migrated as a result of flash floods and floods as in 2013. Indeed, storm hazard-induced areas were expanding. Ky Loi, where 56% of the households had to move, was most affected by the storms. Flash floods caused the displacement of fewer households in Ky Xuan. In the Ky Phu commune, 123 households had to move, and Ky Khang decreased by 66 families (KAG 2014).

The local authorities of Ky Anh paid more attention to limit and overcome the effects of flash floods and other natural hazards than to the whereabouts of the affected households. In 2008, some of the communes were not or to a limited degree affected by the floods. However, in 2013, a number of households had to be relocated: 199 households were affected in Ky Phuong, 70 households in Ky Nam, and 87% of all households had to be relocated in Ky Ninh. In Ky Loi, more households had to be relocated from the flooded area in 2008; however, this number was less in 2013 (Fig. 3.10).

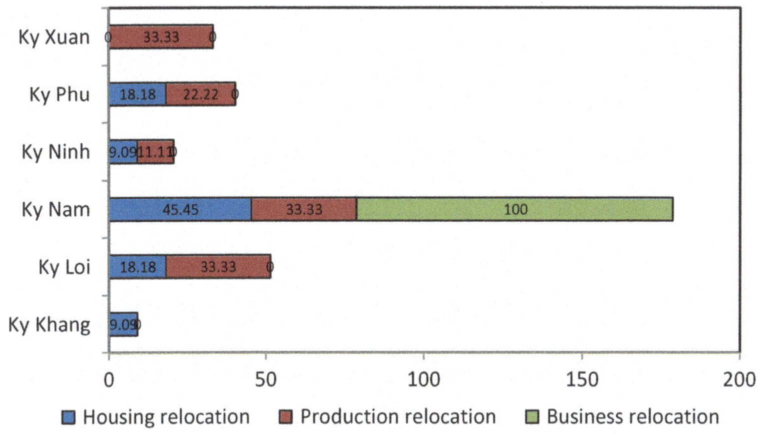

Fig. 3.10 Migration as a result flash floods and floods of 2008 and 2013. Horizontal axis indicates number of migrant households. (Source: KAG 2009, 2014)

In Ky Ninh, almost half of affected households were relocated. Thirty-three percent of the households in Ky Loi were displaced and had to relocate. The migration is associated with the nature of the hazards, in particular with floods and storms. Ky Loi and Ky Ninh have the highest percentage of relocated migrants (12.82%). The Ky Khang and Ky Phu communes have 11% and 9% households relocated, respectively, which also points to natural hazards as the driver of migration. These communes face sea level rise and coastal erosion, in particular in Ky Phu, where sea moved inland driving families away from the coast to inland. The Ky Phuong and Ky Nam communes show 11.29% and 3.78% of household migration. Some villages in these communes host a significant proportion of migrants mainly in the Minh Hue village (Ky Phuong) and the Ba Dong village (Ky Nam), which were built as resettlement neighborhoods for households from the Dong Yen village (Ky Loi). Households relocated twice: once to the Phu Hai village (Ky Phu) and a second time to the Dong Yen village (Ky Loi).

Three major types of relocation prevail along the coast of Ky Anh: housing, production, and business relocations. Migrants driven by housing problems constitute the largest fraction. Each village has its specific causes of migration: Ky Nam shows the highest rate (45%) because of the relocation policy, 33% is the result of production relocation, and 100% is related to business relocation. Migration rates in Ky Xuan, Ky Loi, Ky Ninh, and Ky Phu are highly driven by production (Fig. 3.11).

Local authorities have to inform timely about the expected natural hazards. Only in this way people will be able to prepare, moving to safe places while reducing the damage to their houses. However, the support by the local government proved very limited. On local support after natural hazards, 42.8% of the respondents replied that they did not receive any support while they suffered damage; 58.2% indicated they received (major) support in-kind. The local support is limited and insufficient to start up their income activities again (Fig. 3.12).

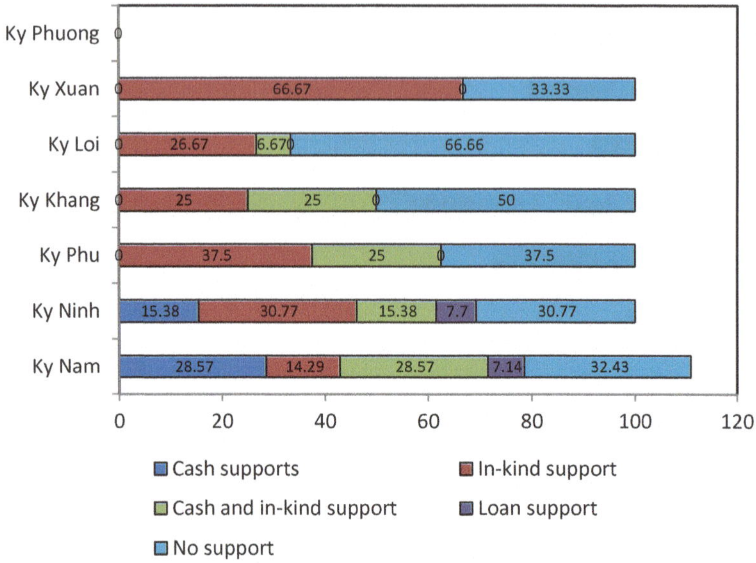

Fig. 3.11 Local perception of driving forces of migration in Ky Anh during 2008–2013. Horizontal axis indicates number of migrant households. Horizontal axis indicates the percentage of respondents reported. (Source: household questionnaire in 2013)

Fig. 3.12 Local perception of post-hazard support in Ky Anh during 2008–2013. Horizontal axis indicates the percentage of respondents reported. (Source: household questionnaire in 2013)

Along the Ky Anh coast, natural hazards have both direct impacts on migration and indirect impacts on land use. These factors have a critical role on the livelihood of the local population in terms of housing, production, or business relocation. Therefore, an effective solution to improve the local adaptation capacity on climate change adaptation is adjusting both local land use methods and land use planning. Because land degradation is a driving force to change livelihoods, adaptation activities depend on new land use methods by local people. Proposing more effective land use methods and land use planning is most necessary to help local people adapting successfully to climate change which is increasing by present in this area.

3.4 Conclusion

The result of this short chapter points to a set of relations between the hazards and the daily life in the coastal communes:

- People face problems with water supply, LULC changes, and coastal erosion. These are general problems faced by numerous coastal communes in Vietnam. The area is subject to progressing urbanization, mainly by converting agricultural land in areas for house and building construction. Of notice is also the increasing surface with planted (forest) trees.
- Along the coastal communes of Ky Anh, the surface of the beaches is subject to a dynamic process in which erosion dominates accretion during 40 years. The extent of the erosion varies from a few meters in Ky Khang to 120 meters in Ky Xuan during 1973–2013. The processes underlying the erosion are complex and multifactorial. Part of natural factors is the storms of which the effect is enhanced by sea level rise (20–25 cm during the 2003–2023 decade).
- Two of the communes deal with a significant gross out-migration of in total 4000 (young) people from 1400 families. Although a relation between migration, on the one hand side, and storms and flash floods exists, also migration is multifactorial and influenced by housing, agricultural yields, and job opportunities. It is important taking notice of the perception of the inhabitants that authorities provided insufficient post-impact support.

Uncertainties of study results in this chapter are associated with the factors people perceive as least important. Overall floods and to a lesser extent sea level rise are hardly mentioned by the interviewees. In particular for the items considered of limited importance (livestock, income), this might be the result of bias. This study ranks effects of the hazards according to people's perception based on mean values in a four-point scale of Likert. The Likert scale is widely used in this type of research and is considered an effective tool in many disciplines. However, the Likert scale is known having main disadvantages: more complex relationships are hard to assess in this way, while the respondents are unaware of this limitation.

Chapter 4
Gross Costs and Impact on Local Livelihood Capitals of Tropical Storms Along the Ky Anh Coast

Abstract This chapter is complementary to the previous one. It analyzes the impacts of storms on people's life and on the management by the local authorities more in detail. The benchmark for this chapter is the devastating storm of September 2013. Two main aspects are addressed: the perception of people and the local authorities in coastal communities of the Ky Anh district and the second part monetarizing the direct gross cost of the damage.

The local perception part of the chapter provides the results of three core aspect: (i) The perception of local households is inventoried using a questionnaire which was completed household by household during the field work in the Ky Anh district. In particular, the details of the effects of the storm weather on crops, farm animals, aquaculture, and property are addressed. (ii) The perception of the inhabitants is complemented with the elements provided by the local authorities. (iii) The perception of the stakeholders in the communes of Ky Anh is added to the two previous elements. To this end, the seven coastal communities are classified in three groups according to their main character: ruralizing, urbanizing, and industrializing.

The second part of this chapter monetarizes the impacts for the households of the September 2013 storm. Also this section addresses aspects of the issue: the quantification of the gross costs on three livelihood capitals (natural, physical, and financial) and the direct costs of the benchmark storm which are put in perspective by a comparison of the cost of other storms in the selected communities.

This section concludes with an ARIMA model trend analysis of the gross damage costs during 10 years. The analysis is based on data from the past 7 years and looks for the situation during the period 2014–2019.

Keywords Local perception · Stakeholders · Gross costs · Livelihood capitals · Tropical storm · Trend analysis · Ky Anh coast

A. T. Nguyen, L. Hens, *Human Ecology of Climate Change Hazards in Vietnam*, Springer Climate, https://doi.org/10.1007/978-3-319-94917-8_4

4.1 Local Perception and Sectorial Impact of Climate Change Hazards

4.1.1 Perception of Local Households

Table 4.1 shows the results of the household questionnaires in the three coastal communes of the Ky Anh district: Ky Xuan, Ky Khang, and Ky Phu. The Likert wMean values for storms, floods, sea level rise, and heavy rains are provided for the three communes and for the five economic sectors under study. The most striking finding is that in spite of the floods and the extreme weather, a small minority of the households in Ky Phu and Ky Khang mentions impacts on livestock and income that can be linked to the storms. Moreover, the results show that storm is perceived as the most threatening situation. (Suddenly arising) floods are considered the least impacting phenomenon. This is related to the location of the fields, behind the dunes.

Sea level rise and the associated floods are considered impacts of equal importance in Ky Phu. In contrast, sea level rise is considered of marginal to no importance in the two other communes. The perception on livestock is influenced by the number of animals the interviewed families lost. A similar interpretation applies to the finding that (sudden) floods are considered of marginal importance in Ky Xuan and Ky Phu, but not in Ky Khang. Storms, sea level rise, and heavy rains are considered the most impacting factors on aquaculture. (Sudden) floods are considered not or most marginally affecting shrimp, shells, and fish aquaculture and were unanimously reported having the least impact on the household income. They

Table 4.1 Perceived impacts of natural hazards on local communities along the Ky Anh coast in the period 2008–2013

		Likert wMean value			
Sector	Commune	Storm	(Sudden) flood	Sea level rise	Heavy rain
1. Crops	Ky Xuan	1.30	0.40	0.70	0.70
	Ky Phu	1.00	0.88	1.00	0.56
	Ky Khang	1.00	0.43	0.71	0.57
2. Livestock	Ky Xuan	0.37	0.19	0.11	0.7
	Ky Phu	0.21	0.25	0.25	0.13
	Ky Khang	0.05	0.00	0.00	0.19
3. Aquaculture	Ky Xuan	1.44	0.00	0.22	0.33
	Ky Phu	2.00	0.00	1.62	1.13
	Ky Khang	2.57	0.71	2.14	0.86
4. Income	Ky Xuan	0.50	0.13	0.28	0.27
	Ky Phu	0.71	0.37	0.82	0.46
	Ky Khang	0.79	0.31	0.66	0.34
5. Property	Ky Xuan	0.35	0.09	0.22	0.22
	Ky Phu	0.87	0.66	0.97	0.68
	Ky Khang	0.76	0.41	0.91	0.27

Source: survey questionnaire 2013

hardly affect the property value of the families. Sea level rise and storms are perceived in Ky Phu and Ky Khang as impacting most the property of the households. The income affection varies by the commune.

4.1.2 Observations and Opinion of Local Governments

Standardized interviews allow inventorying the observations and opinions by local authorities on the impact of storms and floods in the study area. The September 2013 storm was the reference point. As shown in Table 4.2, it affected people, houses, local infrastructure, and the access to fresh water. The Ky Khang commune was affected more than the two neighboring communes. The damage is quantified and the economic impact is calculated. Both the total direct costs and the split over the livelihood capitals are included. The human ecological approach allows dealing with the uncertainties characterizing the environmental, well-being, economic, and policy aspects of these storms.

The opinion of the local government to overcome the risks of the storm effects was inventoried during meetings and based on completed questionnaires. The questionnaires were about significant indicators which were already used by the Vietnam's National Strategy on Climate Change and the National Green Growth Strategy during the period of 2010–2020. Next to climate change adaptation and hazard prevention indicators, green growth is addressed. Green growth is used in Vietnam's policy reply to Rio + 20, but the concept covers a "glossy," political terminology. On the other hand, it is rather well defined in the official documents and well understood by politicians and administrators.

Table 4.3 shows the results of the opinion of the local government representatives in the Ky Xuan, Ky Khang, and Ky Phu. The government of the Ky Xuan commune optioned for solutions of 1.1–1.5, 2.2, and 2.3 targeted to specific actions including planting trees to stabilize the sand and building gulf breakers, lake dams, and water pumping stations along the river. Policy responses entail integrating hazard prevention in transportation plans (harbors, constructing a military road along the coast, building standards resilient to the sea level rise scenarios), and land use planning (replacing paddy rice by drought resistant plants).

The officials of Ky Khang optioned for items 1.1, 1.5, and 2.2. They suggest developing irrigation systems, upgrading the drainage system, planting protection forests along the coast, and constructing sanitation – pollution spreading prevention dams; they also intend developing policies integrating hazard prevention in the new rural development planning, evacuating sites during the storm season, and plans to convert rice paddies into peanut fields.

The officials in the Ky Phu commune prefer items 1.1, 1.4, 1.5, and 2.2. They would like acting on planting protection forests along the beaches, building gulf breakers, integrating hazard prevention in relocation plans for houses close to the sea, new rural development planning (building new flood protection facilities, increasing the height of the road), and land use planning (new residential areas, planting protection forests).

Table 4.2 Damage caused by the September 2013 storm along the Ky Anh coast

	Damages	Units	Ky Xuan	Ky Khang	Ky Phu
1.	*Directly affected people*				
	Affected households	Number	0	1031	0
	Affected people	Number	0	4125	0
2.	*Housing*				
	Collapsed buildings	Number	11	0	0
	Damaged houses	Number	735	1031	377
3.	*Social infrastructure*				
	Damaged schools	Number	3	12	5
	Damaged dormitories	Number	4	4	10
	Damaged hospitals/medical centers	Number	1	2	1
	Damaged cultural facilities	Number	9	0	2
	Damaged offices	Number	3	0	0
	Damaged markets	Number	0	1	1
	Damaged utilities	Number	30	25	5
	Damaged electricity cables	Meter	600	5924	2000
4.	*Agroforestry*				
	Damaged crop area	Hectare	14	13	10
	Damaged fruit tree area	Hectare	0	2	2
	Damaged forest area	Hectare	132	36	77
	Damaged seedlings	Thousand	0	50	0
	Dead livestock	Number	4	0	16
	Dead poultry	Number	1800	0	414
	Permanent by eroded crop land	Hectare	0.5	0	0
	Permanent by eroded settlement land	Hectare	0	0	2
5.	*Irrigation infrastructure*				
	Damaged dikes	Meter	370	1275	100
	Other irrigation facilities	Number	12	0	3
6.	*Transport infrastructure*				
	Damaged bridges	Number	0	2	4
	Eroded national and provincial road	Meter	0	0	120
	Flooded national and provincial road	Meter	0	0	300
	Eroded rural road	Meter	0	0	1000
	Damaged boats	Number	0	1	2
7.	*Industrial infrastructure*				
	Damaged utilities	Number	30	25	5
	Damaged electricity cables	Meters	600	5924	2000
8.	*Drinking water and sanitation*				
	Households deprived of clean water	Numbers	1200	0	0
	Affected wells	Numbers	0	0	2

Source: data from local authorities, 2013

Table 4.3 Opinion of the local government representatives on climate change adaptation, hazard prevention, and green growth

Selected policy options	Ky Xuan	Ky Khang	Ky Phu
1. *Climate change adaptation solutions*			
1.1. Proactively coping with natural hazards	•••	•••	•••
1.2. Guaranteeing food security and water resources	•••	••	••
1.3. Actively responding to sea level rise	•••	••	••
1.4. Protecting and developing forests	•••	••	•••
1.5. More explicit policies (on climate change adaptation) by local authorities	•••	•••	•••
1.6. Building climate change resilient communities	••	••	••
1.7. Applying advanced technologies	•	•	•
1.8. Promoting international cooperation	•	•	•
1.9. Diversify financial resources and effective investment	•	•	••
2. *"Green growth" solutions*			
2.1. Develop sustainable organic agriculture	••	•	•
2.2. Review and revise master plans	•••	•••	•••
2.3. Develop key sustainable infrastructure	•••	••	••
2.4. Develop new green villages	••	•	•
2.5. International cooperation	•	•	•

(Of which •••, strong support (all authority representatives in the panel agree that these solutions support adaptation and mitigation climate change and hazard); ••, medium support (only partial, but still a majority consensus among the panelists on the solutions); •, limited support (a minority of the authority representatives suggests acting on this item))

The district government of Ky Anh coordinates the communes. They act on annual communication and involving stakeholders on storm control. They inform people on storm prevention in March and support recovery actions after natural hazards.

4.1.3 Stakeholders' Opinion

The three communes that were seriously affected by climate change and natural hazards might particularly take advantage of a tailored land use planning policy. Ky Phu, Ky Khang, and Ky Xuan (group 1) are rural communes, of which the dominant existing land use is rural settlement and agriculture. The inhabitants were severely impacted by flash floods and sea level rise. Ky Nam and Ky Ninh (group 2) are rapidly urbanizing rural communes, of which the dominant planned land use is service, tourism, and urban expansion. This is likely strongly impacted by heavy rain, storms, sea level rise, and floods. The Ky Phuong and Ky Loi communes (group 3) face industrialization. The dominant existing land use provides space to industry in the Vung Ang EZ. This area is strongly impacted by storms and heavy rains which cause danger to both local residents and infrastructures (Table 4.4).

Table 4.4 Stakeholder's opinions on climate change adaptation concerning land use and land use planning along the Ky Anh coast

Coastal communes	Climate change drive	Household perception on local land use methods	Authority's opinion on an integrated land use planning
1. Rural communes: Ky Phu, Ky Khang, and Ky Xuan	Sea level rise and coastal erosion cause the loss of residential land	Planting more coast protection forests, reuse land left behind setbacks for the mainland	Incurred households will be arranged housing land patches to live alternating in the 8 villages, except Phu Hai village (Ky Phu) which has a long coast, high population density, and serious coastal erosion Planning new residential areas and expanding planted forests (Ky Xuan) Planning a new residential area along the roads of the Dong Tien village (Ky Khang).
	Flash floods and landslides cause the loss of houses in Ky Xuan and Ky Khang	Relocating houses inland to a safer place	
2. Urbanizing communes: Ky Nam and Ky Ninh	Heavy rain, storms, sea level rise, and mediated floods regularly displace locals	Arrange a new residential area in the hills to avoid floods; plan land for new forests protecting the hinterland Look for solutions to mitigate the impact of floods as houses on the hills away from the flooded area; build publish houses which are not affected by the floods, both for cultural activities and shelter; move assets from households when floods appear suddenly; plan new lakes to drain water; build high dikes and dams to limit the effects of tides and sea level rise	Build resettlement areas in Ba Dong village (Ky Nam) and in Thang Loi village (Ky Loi)
3. Industrializing communes: Ky Phuong and Ky Loi	Storms, cyclones, and heavy rains provide a danger to local people	Recovered land for building industrial zones, and sea ports Plan resettlement areas in other communes	Build resettlement facilities in the Minh Hue village (Ky Phuong) for households who lost their land to build industrial zones

Source: networks between stakeholders, 2013

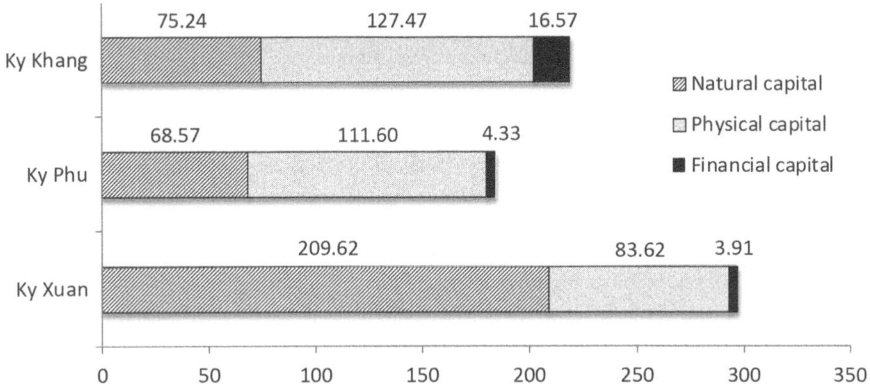

Fig. 4.1 September 2013 storm damage by livelihood capital (thousand $US)

4.2 Valuation of Tropical Storm Damage

4.2.1 Gross Costs and Impact on Livelihood Capitals of the September 2013 Storm

The September 2013 storm caused more damage to local livelihoods than any other storm during recent years. Figure 4.1 shows the cost estimations of the damage to different capitals impacted by the storm. Most sentence on defining the considered capitals. Ky Xuan commune was most affected, followed by the Ky Khang and Ky Phu communes. The damage in the Ky Xuan commune was estimated at about 297.14 thousand $US, while the corresponding figure for Ky Phu were 184.5 thousand $US. In general, the physical and the natural capitals are most affected by the storm in all three communes. In Ky Phu, the physical capital represented 60.5% of the total cost, while in the Ky Khang and Ky Xuan, it accounted for about 58 and 28%, respectively. The natural capital contributed about 70.5% of the total damage in Ky Xuan, while in Ky Phu and Ky Khang, it accounted for 37 and 34%, respectively. The zero values for both the human and the social capitals are likely underestimations of the reality. The September 2013 storm, no doubt, contributed to out-migration of mainly young people in all three villages. However, these latent, longer-term effects are not reflected by the figures provided.

4.2.2 Costs of the September 2013 Storm Compared with Other Storms During 2008–2013

Impacts of storms and related natural hazards on the livelihood capitals are different in the three communes and change yearly during the period 2008–2013. Figure 4.2 shows that livelihood damage due to storms increases sharply in 2010 and 2013,

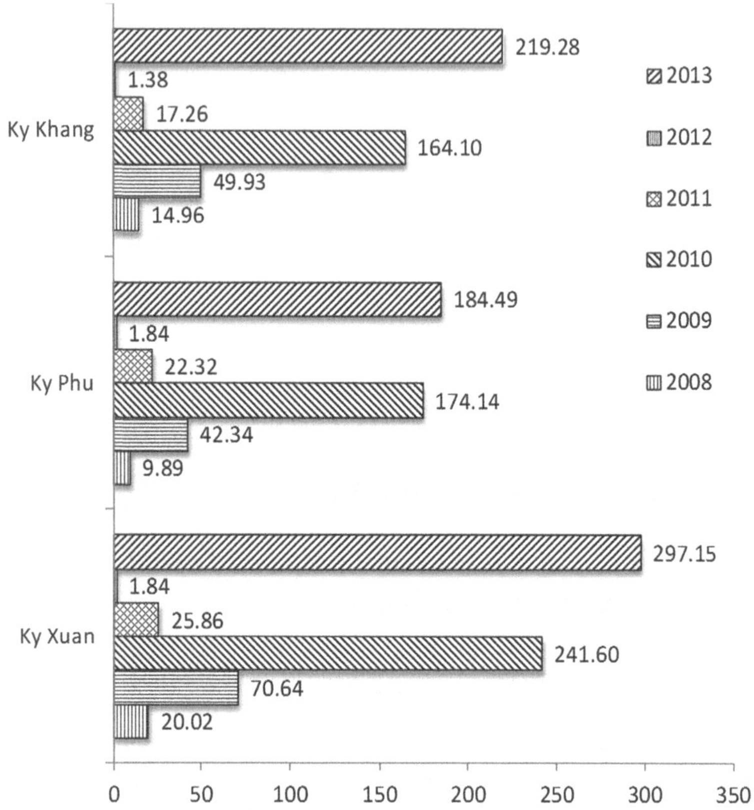

Fig. 4.2 Hazard damage by livelihood in the coastal communes of Ky Anh during the period 2008–2013 (thousand $US)

when this area was directly impacted by storms and accompanying floods. Ky Phu experienced the least damage while Ky Xuan was affected most. In Ky Xuan, people are mainly living on agriculture, forestry, and fishery, whereas Ky Phu has a more diversified geographical and economic basis, including tourism. This explains why the costs are higher in Ky Xuan than in both other communes. In general, the main impacts of storms and related hazards are on aquaculture ponds, dikes, aquatic/sea products, crops and livestock, soil erosion of the fields, destruction of stables, roads and irrigation infrastructure, and degradation of the water quality. The yearly variation documented in Fig. 4.2 illustrates difficulties with forecasts on yearly budgets alleviating storm damage. Budgets are only consistent if they are considered over longer (e.g., 5–10 year) periods.

Table 4.5 shows that the total damage in the communities caused by the tropical storms during the period 2008–2013 is estimated at 1.56 million $US. The figure includes the gross cost of damage which amounts to 0.26 million $US per year. The gross figure entails:

Table 4.5 Cost of damage caused by storms and related hazards to livelihood capitals in the study area during the period 2008–2013 (thousand $US)

Commune	Year	Natural capital	Physical capital	Financial capital	Livelihood related damage by year
Ky Xuan	2008	0.00	19.56	0.46	20.02
	2009	63.05	7.13	0.46	70.64
	2010	192.82	46.71	2.07	241.60
	2011	1.47	23.65	0.74	25.86
	2012	1.15	0.69	0.00	1.84
	2013	209.62	83.62	3.91	297.15
Total damage		**468.11**	**181.36**	**7.64**	**657.11**
Ky Phu	2008	0.00	9.66	0.23	9.89
	2009	37.51	4.37	0.46	42.34
	2010	138.29	34.47	1.38	174.14
	2011	1.61	20.02	0.69	22.32
	2012	1.15	0.69	0.00	1.84
	2013	68.57	111.60	4.33	184.49
Total damage		**247.12**	**180.81**	**7.13**	**435.02**
Ky Khang	2008	0.23	14.27	0.46	14.96
	2009	44.41	5.29	0.23	49.93
	2010	130.69	32.03	1.38	164.10
	2011	1.38	15.42	0.46	17.26
	2012	0.00	1.38	0.00	1.38
	2013	75.24	127.47	16.57	219.28
Total damage		**251.96**	**195.86**	**19.10**	**466.91**

- *Agriculture:* Total crop losses were estimated at 10 thousand tons; merely rice were destroyed; vegetable losses were estimated at about 196.5 thousand $US during 6 years.
- *Irrigation system:* Heavy rains over large areas resulted in major floods and swollen rivers. The floods damage the irrigation facilities and dams. Heavy rains also increase the water level in the lakes behind the dams; in case the water exceeds the designed capacity, this breaks dams and causes flash floods downstream. The total damage on irrigation facilities is estimated at 598.25 thousand $US.
- *Transportation system:* Heavy rain and flood cause damage to roads, bridges, and dams; they contribute to landslides and sedimentation. These costs are estimated at about 243.9 thousand $US.
- *Housing:* Three types of houses exist in the study area – permanent, semipermanent, and temporary constructions in bamboo. The total damage over 6 years amounts to 874.4 $US, most of which are collapsed roofs (accounting for 61%), destroyed houses (31%), and flooded residences (8%); the semipermanent and permanent buildings were affected most (about 42% of total damaged households).

- *Infrastructure:* Each of the storms and floods caused significant damage to schools, local hospitals, warehouses, office buildings, roads, communications systems, and electricity networks. Most of them are not concrete buildings and thus vulnerable. After each calamity, all activities are affected. They are interrupted; occasionally it took a long time before they were repaired and resumed their activities. Other infrastructure damage includes industrial buildings, ports, farms, and stations.

Of the three coastal communes, Ky Xuan faced the highest damage bill with 657.11 thousand $US in total during 6 years. The damage bills of Ky Khang and Ky Phu accounted for 466.91 and 435.02 thousand $US, respectively. The physical capital in Ky Khang was most affected with an estimated total 195.86 thousand $US. This paid for houses, schools, hospitals, and infrastructure. Also the financial capital was affected in Ky Khang (the bills showed 19.10 thousand $US during 6 years). The figure for the natural capital in Ky Xuan is 468.1 thousand $US, of which damage to rice paddies, cash crops, vegetables, animals, seafood, and forests was the heaviest burden. Damaged dikes enhance the effects of the hazards, in particular these of floods. Other influencing factors include the poor infrastructure and the substandard condition of private houses and stables for livestock.

4.2.3 Trend and Cost Analysis of Tropical Storms

The 2008–2013 data on storm damage can be used calculating a trend and providing a forecast. In the selected coastal communes, the costs increase over time. Figure 4.3 shows the value of the damage prediction for the period 2014–2019 as calculated with the ARIMA model. This model does not calculate confidence limits. The remark is however most significant as the increase is influenced by the high values for 2010 and 2013. The values for the period 2014–2019 show a stable trend. The predicted damage cost decreased in each commune during the period 2014–2016. In 2017, the monetary value of the damage is estimated at about 138 thousand $US in Ky Xuan. The figures for Ky Phu and Ky Khang are lower: about 92 thousand $US. During 2017–2019, the damage value in Ky Xuan is expected to decrease to about 70 thousand $US, and the corresponding figure in Ky Phu and Ky Khang is about 46 thousand $US.

Figure 4.4 shows the results of the cost analysis for the forecasted damage taking into account an interest rate of 7% (this percentage is the interest rate of bonds emitted by the Vietnamese government). The 7% interest rate is compared to the 11% interest rate, which is the interest of the current loans at the Vietnam Commercial Bank for Foreign Trade (Vietcombank). The results of the cumulated costs show increasing values in the three communes. The bill is heaviest in Ky Xuan, where the cumulated cost over the 6 years approximates 600,000 $US, about 200,000 $US, above the cost in Ky Phu and Ky Khang (at an interest rate of 7%). Using however an 11% interest rate, the resulting difference is smaller because higher interest rates result in more limited outgoing cash flows.

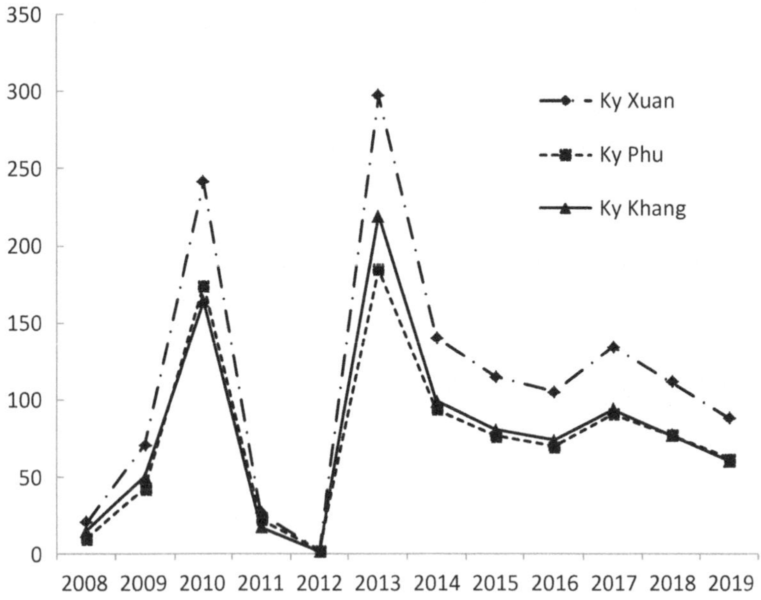

Fig. 4.3 Storm-induced damage observed values and forecasts using ARIMA (1000 $US)

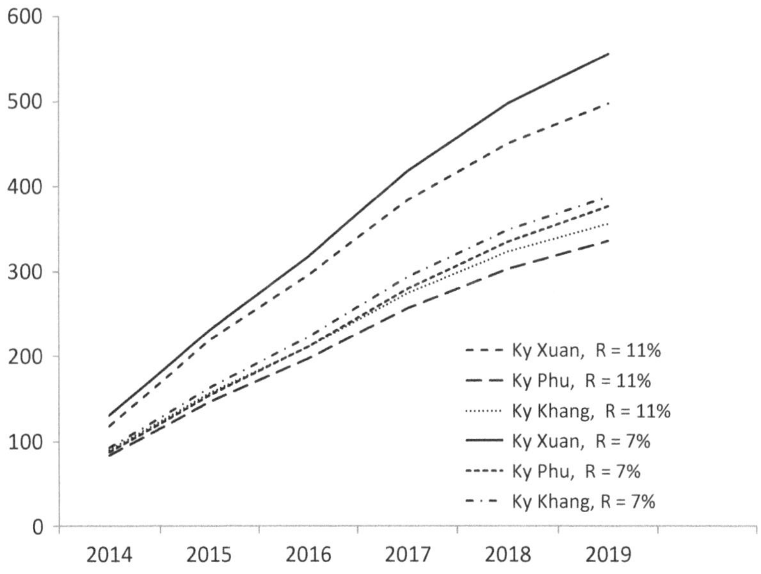

Fig. 4.4 Accumulated cost for forecasted damage during the period 2014–2019 (thousand $US) (in which, R is the interest rate)

4.3 Conclusion

The results of the research synoptically provided by this chapter (for more details, see references) point to interesting conclusions:

- Among the different types of hazards (including among others sea level rise and flash floods), the local inhabitants assess storms as the most impacting on their life. Impacts on aquaculture and the yield of crops are of primary importance to them. Significant variation exists on these gross costs among the villages, which shows that no standard pattern of harmful results is detected by this study.
- Alleviating the impacts of the storms, local authorities focus on biotechnical interventions and policy – management measures. Their priorities are on planting trees, planning, and policies to proactively cope with the problems. International cooperation, awareness raising and education on climate changes rank only in a second order on their agenda.
- Although good theoretical reasons for this position of the authorities exist, these priorities only partially coincide with these of the stakeholders in the villages. In the rural communities, these latter put emphasis on planting trees, protecting the vulnerable neighborhoods, and providing new houses in inland areas which are less prone to inundation. In the urbanizing and industrializing communes, the focus is also on planning safer locations away from the floods.
- The analysis of the gross costs abating the damage shows that primarily the natural and physical capital assets are affected. Authority budgets alleviating the ex post effects of the storms vary: they can only be planned on a midterm basis of 5–10 years.

Overall, the chapter shows storms have devastating impacts not only on the houses and the physical environment but also on the perception of people living in the affected areas. A first estimate of the costs of this damage is presented. The estimate goes beyond the direct payments by households and (local) authorities. Instead three types of capital which are relevant in a more comprehensive context for the households are assessed. The results show that first of all the physical capital is affected. Moreover, as storms occur infrequently, significant differences exist over the years and among the most affected neighborhoods. The chapter provides a first estimate of the budget authorities need to set aside compensating for storm costs during the years to come.

Nevertheless, the chapter shows few limitations particularly in its monetization section. First, the study considered mostly on the impacts of storms recently. It is not clear how the risk of "change in climate" in the future has been assessed, and thus, recommendations which are relevant to climate change are not much unconvincing. Secondly, a set of uncertainties in monetization section includes:

- The estimations of the costs caused by the storms are based on data provided by local authorities and the households. These data mainly cover direct costs and tend externalizing other aspects of the different capitals under study.

- The most significant yearly variations in storm frequency and intensity show that yearly budgets alone (as, e.g., used by authorities) are not the most appropriate instrument dealing with compensations for storm impacts. Medium- and long-term budget instruments (funds or bonds) should be envisaged. Nevertheless, combining the existing data, a medium-term increase in the storm abatement costs is evident.
- The forecasting period is limited until 2019 (6 years). Even then, predictions have a limited confidence. On the other hand, it is the best estimate one might currently provide for a most relevant policy question. As this type of data will accumulate in the future, better, more confident, and reliable forecasts will become available.

Overall, the monetary aspects in this chapter should be considered as preliminary. The research is based on measured data. The more case study data as the ones presented here become available, the more accurate and reliable answers one might provide to the legitimate policy question: "How much do climate change (mediated) effects cost?"

Chapter 5
Climate Change Adaptation of Local Communities Along Heavily Damaged Coasts

Abstract This chapter analyzes how major stakeholders (farmers, fishermen, fish traders, and local authorities) living in the coastal villages of Ky Anh assess climate change-associated hazards. A two-round Delphi approach allows to rank a series of effects and mitigation measures. The results contribute to a vision on local adaptation actions and policy to foster the capacity and the resilience of major local groups.

The chapter is structured in two main parts. It takes off with a description of the two Delphi rounds which were organized on questions and statements which fit in the logic of a drivers-pressure-state-impact-responses (DPSIR) model. After Kendall's test showing a strong coherence in the answers of the stakeholder groups, the data are restructured in vision on priority action for the region.

Keywords Local perception · Stakeholders · Gross costs · Livelihood capitals · Tropical storm · Trend analysis · Ky Anh coast

5.1 Delphi Survey Round 1

5.1.1 Revised Open Questions

The impacts of climate change-associated hazards and adaptation capacity in 20 coastal villages of the Ky Anh district are considered in the section. A combination of the Stakeholder Delphi technique and the DPSIR framework was used. Delphi questionnaires allowed assessing the consensus among the respondents of a stakeholder group. Twenty questions and 20 statements were listed reflecting the DPSIR components. Thirty-six panel members, who were randomly selected from four stakeholder groups which included local authorities, farmers, fishermen, and fish traders, were involved in a two-round Delphi process.

The first round questionnaire entails 20 open questions based on two sources: a comprehensive literature review and information provided during the preliminary stage. It takes about 30 min to complete the questionnaire. Five out of the 20 questions needed modification to improve the clarity and the accessibility for the respondents (Table 5.1). These changes allowed optimizing the questionnaires and adjusting the statements to the objectives. The result was a revised list of "questions" for the panel members.

A. T. Nguyen, L. Hens, *Human Ecology of Climate Change Hazards in Vietnam*, Springer Climate, https://doi.org/10.1007/978-3-319-94917-8_5

Table 5.1 Revised open questions in the Delphi survey round 1

Original questions	Revision
1. The use of scientific terminology, such as "extreme weather," "climate change," or "climate change adaptation," resulted in confusion among the respondents	These terms were replaced by a more correct accessible and more understandable vocabulary, including "floods," "prolonged droughts," and "tides"
2. Fuzzy terms resulted in misinterpretation by the respondents	For instance, "agricultural human activities" changed to "agricultural activities of the local people"
3. To section 1: one question (item 2) was added – "In your opinion, what are the economic impacts apart from agriculture on the quality of the local environment?"	This replies to the finding that also nonagricultural economic sectors have important effects on the climate and the local environment in Ky Anh
4. The impacts of climate change on local people: next to negative effects, climate change also has positive impacts on locals	Item 5 was added: "In your opinion, does the tidal phenomenon that changes seasonal planting have any positive impact?". If the answer is yes, respondents are requested moving to question 6 – Impact "What are these beneficial effects?"
5. Add the question, "What are the specific measures to reduce the negative impacts of climate change?"	Item 5 has been added: "To reduce global warming, it is indicated to invest more in renewable energy. What are these sustainable energy resources?"

5.1.2 Drivers

Two main drivers from agricultural and nonagricultural sectors were mentioned in the questionnaire. Primary causes which adversely affected the environment are the overuse and inadequate application of fertilizers and pesticides (chosen by 9/36 respondents) and the overuse of fertilizers and plant protection chemicals (8/36). These elements were merged and mentioned in statement "S_1.1" used in round 2. Nonagricultural activities also affect the environment. Mining is the most significant one (17/36), followed by transport (6/36). Building hotels and economic development projects along the coast causes mangrove deforestation, resulting in a decrease of both forest and agricultural land. The nonagricultural activities are presented as the statement "S_1.2" in round 2 (Table 5.2).

5.1.3 Pressure

Main pressures on changes in the agricultural production include migration, calamities, climate change, and population growth (Table 5.3). 25/36 respondents indicated migration as the main reason for these problems; 21/36 indicated that calamities and the change of climate were the second most important reason for the pressure on the yields; and 12/36 identified population growth as the third most important reason. The second question asked to identify the populations most affected by calamities and climate change hazards. Most respondents pointed to

Table 5.2 Questions on drivers (D) and reply by the respondents in round 1

Code	Question	Most frequently selected alternative	Number of responses/total number of responses	Symbol used in round 2 statement (S-code)
Q_1.1	What are the main drivers of agriculture adversely affecting the environment?	Overuse and improper use of fertilizers and pesticides	9/36	S_1.1
		Overuse of fertilizers and plant protection chemicals	8/36	
Q_1.2	What are the main drivers of nonagricultural activities adversely affecting the environment?	Mining	17/36	S_1.2
		Transport	6/36	

Table 5.3 Questions on pressures (P) and reply by the respondents in round 1

Code	Questions	Most frequently selected alternative	Number of responses	Symbol of the round 2 statement (S-code)
Q_2.1	What are main pressures of change in agricultural production?	Migration	25/36	S_2.1
		Calamities	21/36	
		Population growth	12/36	
Q_2.2	Which groups of the population were most affected by calamities and climate change hazards?	Farmers, fishermen, and fish traders	36/36	S_2.2
Q_2.3	Which sectors were most affected by calamities and climate change?	Mining industry, processing industry, and agriculture	36/36	S_2.3

farmers, fishermen, and fish traders who appear in the statement "S_2.2" of the round 2. Sectors as the mining industry, the processing industry, and agriculture, which are most affected by calamities and climate change, are considered in statement "S_2.3" of round 2.

5.1.4 State

Five questions and the most frequently selected alternatives describing extreme weather events are listed in Table 5.4. Storm, floods, and prolonged drought occurred more frequently and more destructively in the previous 20 years (18/36) (Question Q_3.1). Climate change-associated hazards include the increasing frequency and

Table 5.4 Questions on state (S) and reply by the respondents in round 1

Code	Questions	Most frequently selected alternative	Number of responses	Symbol of the round 2 statement (S-code)
Q_3.1	Since when do extreme weather events occur more frequently and more destructive during recent years?	Since 10 to 20 years ago	18/36	S_3.1
Q_3.2	What are climate change-associated hazards during the past years?	The frequency and intensity of storms increase	18/36	S_3.2
		Duration of storms, floods, and droughts increases	34/36	
		Extreme weather events cause more damage	35/36	
Q_3.3	To which extent did climate change hazards alter agricultural activities?	Considerable change	19/36	S_3.3
Q_3.4	Which crop was substantially affected by changes of extreme weather events, particularly in the delta (these result in decreased yields)?	Rice	35/36	S_3.4
Q_3.5	Which crop was substantially influenced by extreme weather events, particularly along the coast and in the estuary, and experienced decreased yield?	Peanut	30/36	S_3.5

intensity of storms (18/36). The duration of the storms, floods, and prolonged droughts increased (34/36); extreme weather events happen more often during recent periods (35/36) (Question Q_3.2). Sudden changes in agriculture are driven by climate change (19/36) (Question Q_3.3). Agricultural yields of rice and peanuts are most affected by changeable and extreme weather in the plains, along the coast and in the estuary (Question Q_3.4 and Q_3.5). The most frequent answers are rephrased as statements S_3.1, S_3.2, S_3.3, S_3.4, and S_3.5 for round 2.

5.1.5 *Impacts*

Four questions deal with the impacts of climate change hazards and extreme weather events on the agricultural production (Table 5.5). Natural hazards as more intensive storms than before, floods, and drought impact on the field (32/36) and affect the yields, causing more pests and diseases (29/36) (Question Q_4.1). Extreme weather

Table 5.5 Questions on impacts (I) and reply by the respondents in round 1

Code	Questions	Most frequently selected alternative	Number of responses	Symbol of the round 2 statement (S-code)
Q_4.1	What are the main impacts of climate change hazards (storms, floods, prolonged drought) on agriculture?	Impact on yields	32/36	S_4.1
		Increased risk of pests and diseases	29/36	
Q_4.2	What are the main impacts of extreme weather (higher temperature, intensive storms, and prolonged drought) on agriculture?	Changes in agricultural land	29/36	S_4.2
		Increase in pesticides use	25/36	
Q_4.3	How does sea level rise affect agriculture?	Harmful	20/36	S_4.3
Q_4.4	Does sea level rise have beneficial effects on agriculture?	Local authorities provide support for agriculture	25/36	S_4.4

events as higher temperature, intensive storms, and prolonged drought cause changes on agricultural land (29/36) and increase pesticide use (25/36) (Question Q_4.2). The diverse effects of sea level rise on the yields further increase the concerns of the local authorities (Question Q_4.3 and Q_4.4).

5.1.6 *Response*

The prevailing answers to six questions on responses are shown in Table 5.6. Construction and strengthening dikes and irrigation systems (22/36) and dealing with damage caused by natural disasters (19/36) are considered as the most effective measures implemented by local authorities (Question Q_5.1). Current adaptation measures to alleviate climate change impacts are insufficient (25/36) (Statement Q_5.2). Agricultural activities should be complemented with other economic actions (27/36) (Question Q_5.3). Negative impacts of greenhouse gasses should be reduced by saving energy (Question Q_5.4). Planting more trees (24/36) and investments in better waste management are considered the best measures to alleviate the impact of climate change (15/36) (Question Q_5.5). Regarding renewable energy, wind is preferred by 24/36 respondents, followed by solar energy (19/36 respondents) (Question Q_5.6). The two options were offered in the questions of round 2.

Table 5.6 Questions on responses (R) and reply by the respondents in round 1

Code	Questions	Most frequently selected alternative	Number of responses	Symbol of the round 2 statement (S-code)
Q_5.1	How do local measures help farmers adapt to climate change and reduce the impact of extreme weather?	Construction and upgrading dikes and irrigation systems	22/36	S_5.1
		Providing support to deal with the damage caused by natural hazards	19/36	
Q_5.2	Assess the current adaptation measures to cope with climate change	Inadequate	25/36	S_5.2
Q_5.3	How do you assess diversifying economic activities next to agriculture?	Necessary	27/36	S_5.3
Q_5.4	How to reduce the negative impacts of greenhouse gasses by saving energy?	At home and in daily life	30/36	S_5.4
Q_5.5	What are measures to alleviate the impact of climate change?	Planting trees	24/36	S_5.5
		Improvement of waste management	15/36	
Q_5.6	Which type of renewable energy should be developed?	Wind energy	24/36	S_5.6
		Solar energy	19/36	

5.2 Delphi Survey Round 2

5.2.1 Statement Statistics

Table 5.7 shows the results of each statement in round 2. Mean values indicate the degree to which local people support or prioritize the statements using a 10-point Likert scale. Standard deviations vary over a small range (between 0.0 and 2.0). This indicates that the variation between the responses is limited and that the information provided by the stakeholders is consistent. It also points to a consensus among the panel members on the effects of climate change on their livelihood and about the environmental impacts, even if they are not close neighbors. The quartile value provides an indication of the variation and the dispersion of the data sets. For the different statements, the benchmark scores are almost equal or only slightly divergent. This confirms that the impacts of climate change hazards on the seven coastal communes are perceived in a similar way.

Figure 5.1 represents the mean values of the statements. Statement S_5.2 attracts the highest mean (9.26), reflecting the highest priority for the stakeholders. The mean values of statements S_2.3, S_3.2, and S_4.3 are higher than or equal to 9. Statements with a low score include S_4.4 (5.06/10), S_3.5 (5.75/10), and S_5.4 (6.62/10).

Table 5.7 Descriptive statistics of the responses to the statements of Delphi round 2

Statements	Mean	Minimum	Maximum	Standarddeviation	Variance	Quartiles		
						50%	75%	5%
S_1.1	8.84	5	10	1.08	1.17	9	9	9
S_1.2	8.71	6	10	1.09	1.18	8	9	9
S_2.1	8.88	7	10	0.91	0.83	8	9	0
S_2.2	8.65	3	10	1.32	1.75	8	9	9
S_2.3	9.26	8	10	0.75	0.56	9	9	0
S_3.1	8.65	2	10	1.76	3.08	9	9	9.75
S_3.2	9.00	3	10	1.28	1.64	9	9	0
S_3.3	8.74	7	10	0.79	0.62	8	9	9
S_3.4	8.26	1	10	1.73	2.99	8	9	9
S_3.5	5.79	2	9	1.74	3.02	5	6	7
S_4.1	8.91	7	10	0.83	0.69	9	9	9
S_4.2	8.59	5	10	1.31	1.70	8	9	9
S_4.3	9.15	5	10	0.96	0.92	9	9	0
S_4.4	5.06	1	9	1.70	2.91	4	5	6
S_5.1	6.26	2	10	2.45	6.02	5	7	8
S_5.2	9.26	7	10	0.67	0.44	9	9	0
S_5.3	8.88	4	10	1.30	1.68	9	9	0
S_5.4	6.62	3	9	1.52	2.30	6	7	8
S_5.5	8.88	6	10	1.01	1.02	9	9	9
S_5.6	8.15	3	10	1.60	2.55	7.25	9	9

Fig. 5.1 Mean values in round 2 of the Delphi survey statements

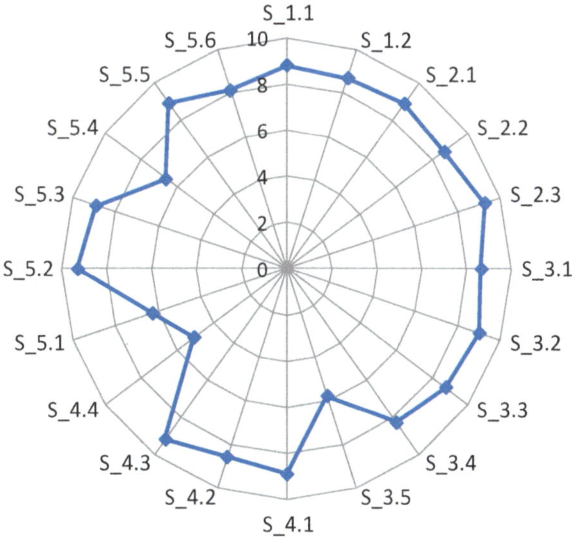

The statistical results of the Delphi round 2 show that:

- *Drivers*: To enhance agricultural yields, local residents tend to use too much pesticide and fertilizer, with significant adverse impacts on the environment. This is illustrated by the high value of the mean (8.84) in statement S_1.1. Statement S_1.2 on the impacts of titanium mining and mangrove deforestation shows that they further degrade the environment (8.71).
- *Pressures:* Calamities, climate change hazards, and population growth are considered main pressures on agricultural yields. Statement S_2.1 realizes a high mean score (8.88). The mean value 8.65 of statement S_2.2 applies to the sectors most affected by calamities and climate change hazards which are the mining industry, agriculture, and the processing industry.
- *State:* Natural hazards and extreme weather show an increasing trend during 20 years. They are addressed by statement S.3.1. Eighty-eight percent of participants selected score 8, 9, or 10, which points to a high level of agreement. The mean of statement S_3.2 is 9.00 (climate phenomena have changed in previous years). Local people consider changes in patterns of storms, floods, and droughts as effects of climate change. More extreme weather events occur over time and cause increasingly more damage. Twenty years ago, the storm season lasted from June to August. However, this season has gradually lengthened until November 2013; periods of drought last longer and have increasingly more impact on agriculture.
- The results of statements S_3.4 and S_3.5 show that the yields of rice and peanuts are most affected by climate change hazards (8.26 and 5.79, respectively). Consequently, local farmers switched to other crops to limit the impacts on their households. Rice and peanuts are traditionally grown in the irrigated plains and river estuaries, making them vulnerable to the effects of sea level rise and storm-induced floods. In 2014, the areas were particularly affected by a long period of cold weather which reduced the rice yield.
- *Impacts:* The mean of statement S_4.1 (impacts of climate change hazards on agricultural production) is high (8.91); the next highest being the mean score is for S_4.2 (8.59). Humid weather brings more pests and diseases and impacts the crops and the livestock, which leads to smaller yields. The current adaptation measure is to overuse pesticides and chemical fertilizers. Statement S_4.3 shows that rising sea levels are known having an adverse impact on the agricultural yields (9.15). However, the rise of the sea level also catalyzes positive aspects, as the increased support for affected farmers by local authorities (5.06) (S_4.4).
- *Responses:* The S_5.1 statement scores medium (6.26). This indicates that adaptation measures by the local authorities such as construction policies, reinforcement of dikes, upgrading irrigation systems, and providing post-disaster support are considered inadequate. Statement S_5.2 (scores the highest mean 9.26) confirms this. Alternative responses advocated by stakeholder members include (1) plant more mangroves and other coastal protection forest, owned and managed by individual families (S_5.3; mean 8.88); (2) invest more in waste management (S_5.4. mean 6.62); (3) use more renewable energy such as wind and solar (S_5.5

Table 5.8 Kendall's W for weights assigned during the second Delphi round

n	Kendall's W	P	Agreement	Confidence in ranks
36	0.681	< 0.001	Strong	High

mean 8.8); and (4) produce electricity from biomass and household waste (S_5.6 mean 8.15). These recommendations are based on the experience of the local people who indicate that current measures are inadequate.

5.2.2 Kendall's W test

The responses by the panel members during the second Delphi round allow calculating Kendall's W of 0.681. As shown in Table 5.8, there is a "strong agreement" of consensus among the panel members, and the "confidence in ranks" is high. Therefore, a third Delphi round is not required.

5.2.3 Summary of Local Adaptation Capacity to Climate Change Hazards

Delphi results in a vision on impacts of and adaptation to climate change hazards with the following dimensions:

- *Drivers*: Overuse of fertilizers and pesticides has a negative impact on the environment. This practice contributes in the long term to land degradation. Moreover, it affects local people's health through exposure to contaminated drinking water and/or food. Nonagricultural activities such as titanium mining inevitably have adverse effects on the environment. Floods in mining areas pollute soil and water. However, the industrial development is also a major factor of extensive land use changes, natural resource uses, and environmental pollutions.
- *Pressures:* Migration, disasters, climate change, and population growth are important factors explaining changes of agricultural yields. Farmers, fishermen, and fish traders suffer directly from rising sea levels and increasingly violent and frequent storms and floods in combination with environmental pollution from the industrial activities in the Vung Ang EZ. As a result, some farmers, fishermen, and fish trader are forced to quit their job.
- *State:* Extreme weather events have increased during the past 25 years, since the 1990s. The panel members confirmed that the intensity of weather events such as storms, higher summer temperatures, and periods of drought have increased during recent years.
- *Impacts*: Agricultural yields, of rice in particular, have fallen due to climate change. However, many other crops are (less) affected by the changing weather.

Fig. 5.2 Planting mangroves to protect the coastline in Ky Nam (Ky Anh). (Source: An Thinh Nguyen, 2013)

The longer storm season also affects the farming season. Farming strategies and techniques had to be adjusted in response to the new situation. Increased rainfall during the rainy season caused changes in agricultural land use. The increased use of pesticides is a cofactor in drought-induced declining agricultural yields.

- *Responses:* The analysis of the responses points to adaptation techniques and systems adopted by local communities in a context of climate changes and damaged coasts. Authorities provided support to local people to alleviate the effects of extreme weather: they assisted locals to evacuate after the storms whenever necessary and supported them to recover from the disaster. The increasing impacts of climate changes on agriculture and the consequent policy decisions explain why agricultural land use changed. Specific mitigating measures include planting forests which protect the coastline (Fig. 5.2) and giving more attention to waste management facilities. Households have been encouraged to use sustainable energy. Local actions are supported by large-scale measures promoting renewable energy sources.

5.3 Conclusion

Internationally, the adaptation capacity of local communities to climate change hazards has emerged as one of the most considerable environmental problems in Asian coastal areas, which are affected by a range of stress conditions and shocks resulting from both climate change hazards and environmental degradation (Cruz et al. 2007; IPCC 2007). Climate change impacts and adaptation are influenced by socioeconomic and environmental elements as poverty, social justice, economic development, and environmental pollution (Eriksen et al. 2011; Colagiuri et al. 2015) and are closely related with local and regional sustainable development initiatives (Njoroge 2014). Only an integrated approach allows understanding this complexity and proposing reasonable adaptation measures (EEA 1995; Doria et al. 2009; Newton and Weichselgartner 2014; Lewison et al. 2016).

The results of this chapter show the relative importance farmers, fisherman, fish traders, and local authorities connect with the changing climatic conditions along the Ky Anh coast. The agricultural production is the area of most concern. The overuse of pesticides and fertilizer originates from an attempt counteracting the gradually reducing yields, in particular of rice and peanuts. These yields are faced by calamities as storms, climate change hazards, and population growth. The current state of these pressures is the result of climate change during the last decades. The current policy to alleviate the situation is assessed as insufficient. Additional action should be taken on planting mangroves, improved waste management, and the use renewable energy resources as wind, solar, and biomass.

These findings show both local and general characteristics. The focus of the different stakeholders on agriculture (and not on marine fisheries and aquaculture) should be noticed. This can be interpreted in a context of immediately visible short-term effects. The increased use of pesticides and fertilizer in a less expected result is also described. Enhanced waste management and the use of renewable energy resources are coherent with the national policy in Vietnam and international recommendations.

Part III
Risks for Nature and Humans in the Uplands

Chapter 6
Impacts and Damage of Climate Change Hazards in the Van Chan Mountain

Abstract The third section of this book is about the climate change effects and risks in the mountains which form the natural border between China and Vietnam. The Van Chan district in this mountainous area was selected as the study area and illustrates the problems in the wider mountain highlands. The district yearly faces cyclones, heavy rains, floods, and also long cold and drought periods. This affects the local agriculture, household income, and livelihood. Therefore the area was selected as a study area which illustrates for the climate change-associated effects in the mountains of the Vietnamese-Sino borderland.

The chapter first addresses the areas at risk. Indexes for landslides and floods allow discriminating places according to the intensity they are prone to these extreme weather effects. A review of the yearly recurring extreme weather situations and long periods of drought and frost and their main effects (floods, landslides, impacts on health and socioeconomic sectors) for the 1999–2015 period shows that the region yearly faces major problems, in particular during the rainy and storm season from July to August in the winter time. An indicative estimation of the direct costs caused by these extreme weather phenomena is provided.

Keywords Climate change hazards · Landslide hazard zone index · Flooding hazard zone index · AHP · Gross cost · Van Chan mountain

6.1 Climate Change

6.1.1 Van Chan Mountain

Van Chan mountain belongs to Yen Bai, a province of Vietnamese northern mountainous region. Yen Bai Province is home to about 800,000 people (2015) and spreads over 6,900 square kilometers (GSO 2016). The province is situated on the Hoang Lien Son mountain ranges with average 600 meters above sea level. The three forest ecosystems in the province include rainforest, subtropical, and temperate forests. In between the hills, one finds fertile valleys formed by the Thao River and the Chay River (Le et al. 2012). The province experiences a tropical monsoon climate with a mean temperature of 20^0C (18–28^0C), which drops below 0^0C during

A. T. Nguyen, L. Hens, *Human Ecology of Climate Change Hazards in Vietnam*, Springer Climate, https://doi.org/10.1007/978-3-319-94917-8_6

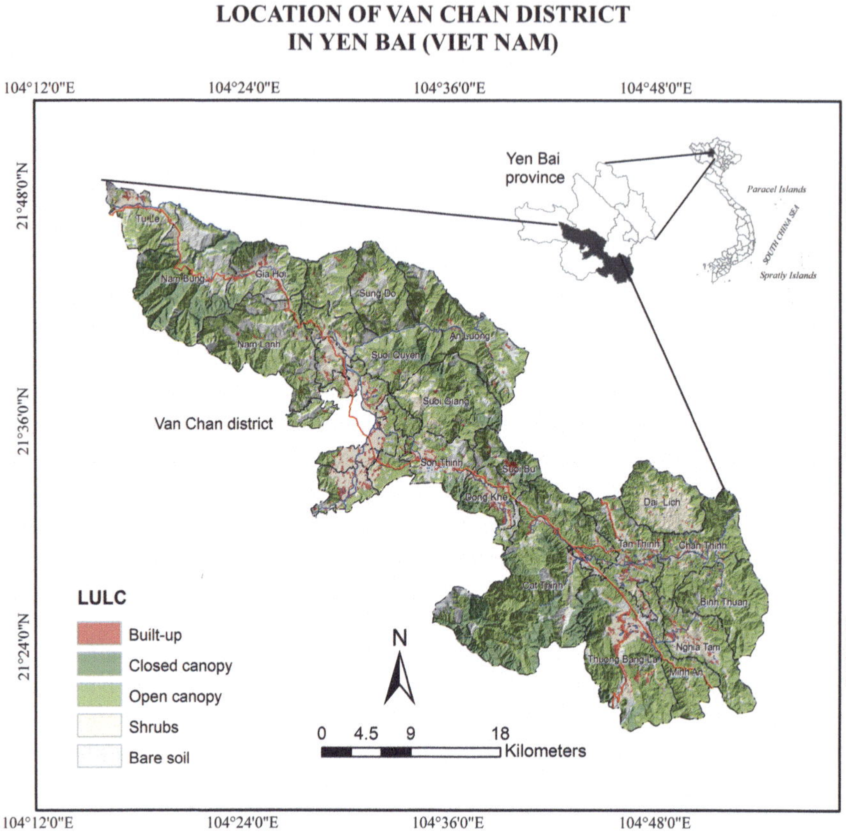

Fig. 6.1 Location of the Van Chan district and Yen Bai Province in the northern mountains of Vietnam

winter time, bringing frost and snow. About 30 ethnic groups live in the province (GSO 2016). Several communes take part in the national program for hunger and poverty alleviation. As shown in Fig. 6.1, Van Chan locates in the southwestern of the Yen Bai Province, about 72 kilometers from the provincial capital. The area covers 1,240 square kilometers, making up about 18% of the province. The landscape is complex because of the interaction between forests, mountains, caves, springs, and flat deltas.

6.1.2 Climate Change

Changes in average annual temperature (^{0}C) and annual rainfall (%) in the study area by 2090 are shown in Figs. 6.2 and 6.3. According to the RCP 4.5 scenario, the average temperature at the beginning of the twenty-first century will increase by

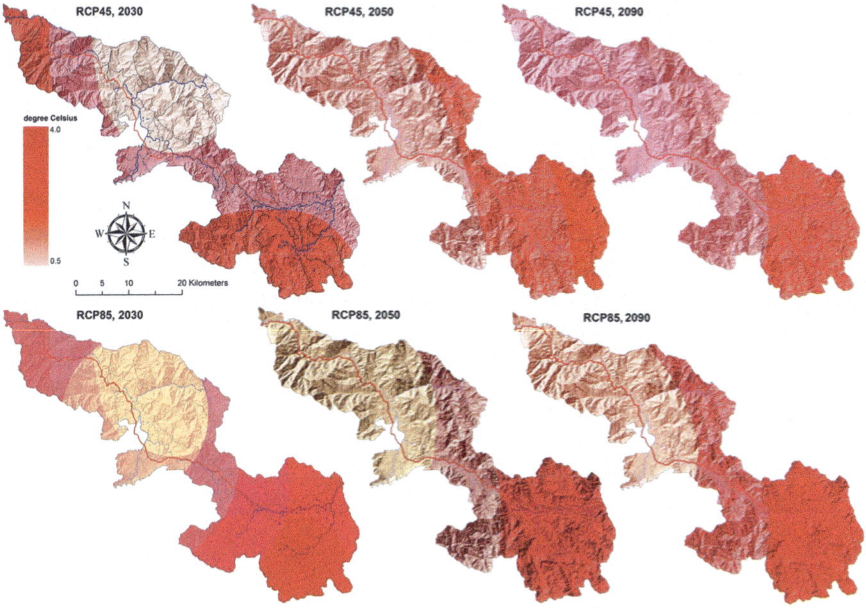

Fig. 6.2 Expected changes in average annual temperature (^{0}C) based on the RCP 4.5 and RCP 8.5 scenarios by 2090 in the Van Chan mountain. (Yen Bai Province, Vietnam)

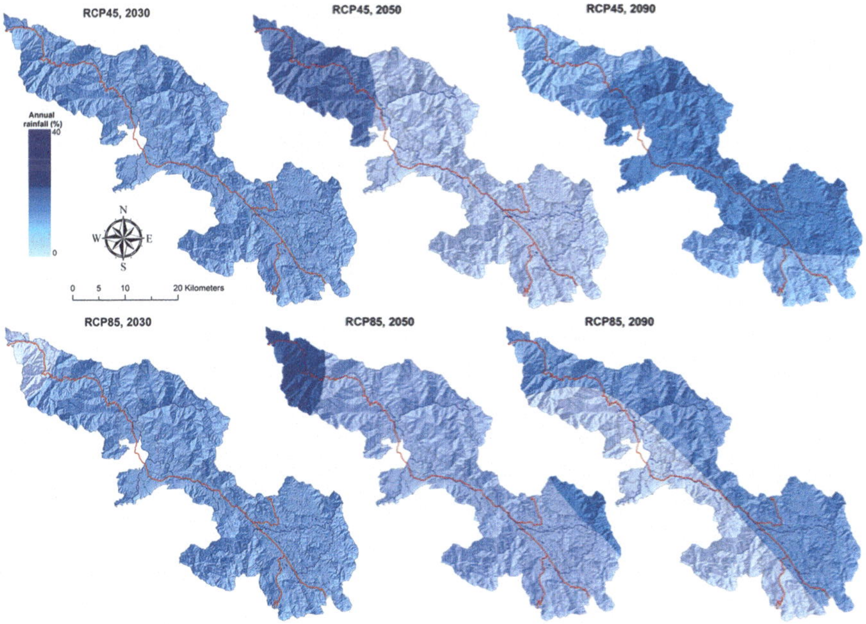

Fig. 6.3 Expected changes in annual rainfall (%) based on the RCP 4.5 and RCP 8.5 scenarios by 2090 in the Van Chan mountain. (Yen Bai Province, Vietnam)

0.59–0.67°C. This figure is expected to increase by 1.6–1.7°C by the middle of the century and by 2.20–2.36°C at the end of the century. According to the RCP 8.5 scenario, the average temperature at the beginning of the century will likely increase by 1.0–1.08°C and by 2.09–2.25°C at the middle of the century and go up to 3.79–3.96°C by the end of the century. Overall, temperatures will likely raise in both RCP 4.5 and RCP 8.5 scenarios (MONRE 2016).

According to RCP 4.5, annual rainfall at the beginning of the century (until 2030) will likely increase by 10–15%; by the mid and the end of the century, the annual rainfall is forecasted to increase by 10–20%. According to the RCP 8.5 scenario, the increasing trend in annual wet deposits will likely be the same as in the RCP 4.5 scenario at the beginning and the mid of the century; and at the end of the century, the annual rainfall will likely increase by 10–25% (MONRE 2016).

6.2 Natural Hazards

6.2.1 Natural Hazard Zones

6.2.1.1 Landslide Hazard

Figure 6.4 shows nine factors in the AHP pair-wise comparison matrix for landslides: elevation (ELE), aspect (slope) (ASP), slope (angle) (SLP), lineament density (LIM), stream density (STE), rock (ROC), land-use land cover (LULC), soil (SOIL), and maximum precipitation (MRE). The AHP result shows that the maximum eigenvalue is 9.253 and the consistency index (CI) is 0.031. The consistency ratio (CR) value below 0.1 indicates a reasonable level of consistency in the pair-wise comparison. This is sufficient to allocate the factor weights. The weight corresponding to slope (SLP) is the most influential (0.306), whereas elevation (ELE) scores lowest (0.020) (Tables 6.1 and 6.2).

Based on the weights of landslide hazard factors, the Landslide Hazard Zone Index (LHZI) values were calculated using the formula:

$$\begin{aligned} \mathrm{LHZI} = & (0.020\,\mathrm{ELE}) + (0.021\,\mathrm{ASP}) + (0.306\,\mathrm{SLP}) + (0.152\,\mathrm{LIM}) + (0.035\,\mathrm{STE}) \\ & + (0.144\,\mathrm{ROC}) + (0.067\,\mathrm{LULC}) + (0.097\,\mathrm{SOIL}) + (0.158\,\mathrm{MRE}) \end{aligned}$$

where ELE is the elevation factor, ASP is the aspect slope factor, SLP is the slope factor, LIM is the lineament density factor, STE is the stream density factor, ROC is the rock factor, LULC is the land-use land cover factor, SOIL is the soil factor, and MRE is the maximum precipitation factor.

For the study area, minimum value (1.0), maximum value (4.0), and a mean value (2.871) of the LHZI were found. The standard deviation was 1.008. The LHZI quantifies the relative hazard of a landslide occurrence: the higher the value of index, the more the area is prone to landslide. These LHZI values were grouped in four classes, which represent four zones in the landslide hazard map: very high

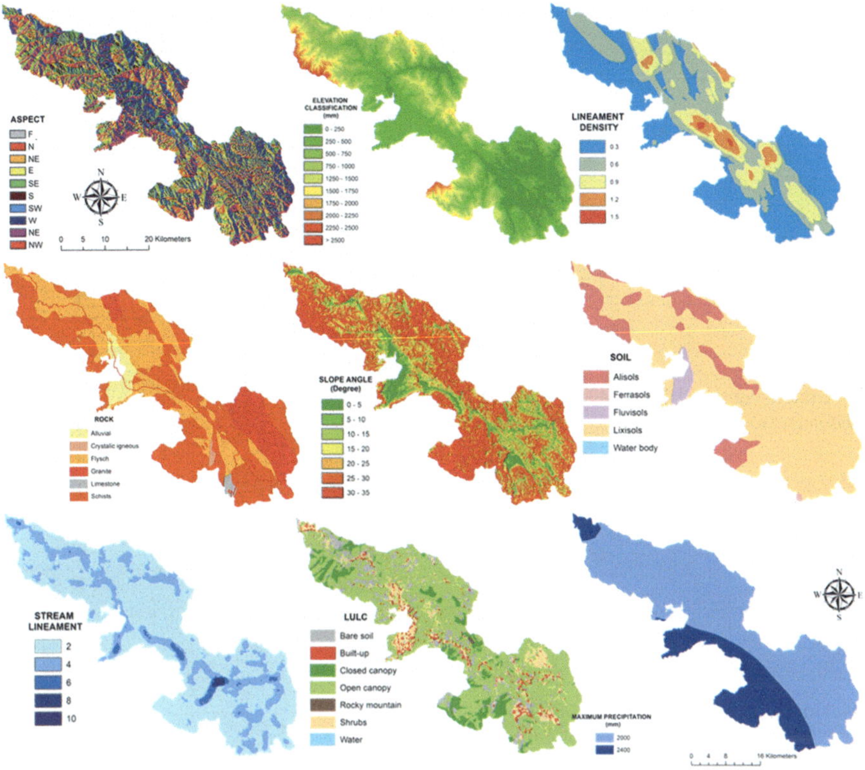

Fig. 6.4 Thematic maps of hazard factors for the Van Chan mountain. (Yen Bai Province, Vietnam)

Table 6.1 AHP pair-wise comparison matrix for landslide

Factors	ELE	ASP	SLP	LIM	STE	ROC	LULC	SOIL	MRE	Weight
ELE	1	1	1/9	1/8	1/2	1/7	1/4	1/6	1/7	0.020
ASP		1	1/8	1/9	1/2	1/8	1/4	1/5	1/7	0.021
SLP			1	3	7	3	5	4	2	0.306
LIM				1	3	1	3	2	1	0.152
STE					1	1/4	1/3	1/4	1/2	0.035
ROC						1	2	2	1	0.144
LULC							1	1/2	1/3	0.067
SOIL								1	1/2	0.097
MRE									1	0.158

Maximum eigenvalue = 9.253; CI = 0.031

hazard, high hazard, medium hazard, and low hazard. Figure 6.5 shows that the Cat Thinh, Thuong Bang La, Dong Khe, Suoi Bu, Son Thinh, Suoi Giang, An Luong, Nam Lanh, Nam Bung, and Tu Le communes show a very high landslide hazard risk. Other communes attract a low landslide hazard figure, Tran Phu, Dai Lich, Binh Thuan, Chan Thinh, Phu Nham, Thanh Luong, and Thach Luong.

Table 6.2 AHP weight and class rating of landslide hazard

Factors	Weight	Class	Rating
Aspect slope (ASP) *[degree]*	0.021	N, NE, NW,S	1
		E: East (67.5–112.5)	2
		W: West (247.5–292.5)	3
		SE: Southeast (112.5–157.5)	4
		SW: Southwest (202.5–247.5)	5
Elevation (ELE) *[meter]*	0.020	<500	1
		500–1000	2
		>1000–1500	3
		>1500–2000	4
		>2000	5
Lineament density (LIM) *[kilometers/sq. kilometer]*	0.152	0.3	1
		0.6	2
		0.9	3
		1.2	4
		1.5	5
Rock type (ROC)	0.144	Alluvial, Limestone	1
		Granite	2
		Crystalic igneous	3
		Flysch	4
		Schists	5
Maximum precipitation (MRE) *[mm]*	0.158	1800	1
		2000	2
		2200	3
		2400	4
		2800	5
Slope (SLP) *[degree]*	0.306	<10	1
		10–20	2
		20–25	3
		25–30	4
		>30	5
Soil type (SOIL)	0.097	Fluvisols	1
		Water body	2
		Alisols	3
		Lixisols	4
		Ferrasols	5
Stream density (STE) *[kilometer/sq. kilometer]*	0.035	2	1
		4	2
		6	3
		8	4
		10	5
Land use land cover (LULC)	0.067	Built up, water	1
		Closed canopy	2
		Open canopy	3
		Shrubs	4
		Bare soil	5

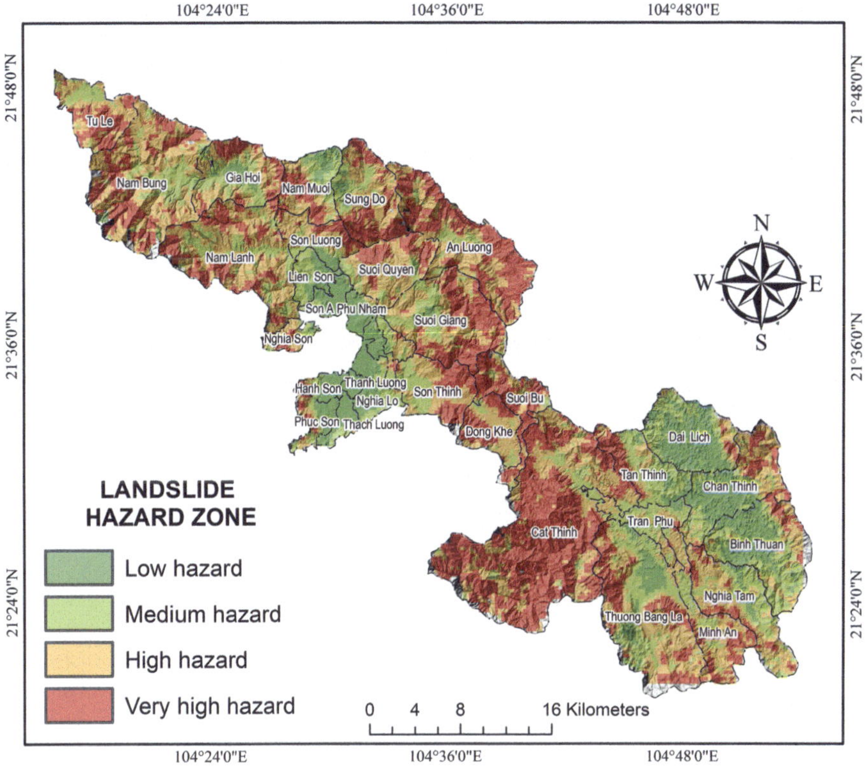

Fig. 6.5 Landslide hazard zones on the Van Chan mountain. (Yen Bai Province, Vietnam)

6.2.1.2 Flooding Hazard

Six factors influencing the flooding hazard were selected to calculate the FHZI: elevation (ELE), slope (SLP), stream density (STE), rock (ROC), soil (SOIL), and maximum precipitation (MAXPRE) (Fig. 6.4). The AHP result shows that the maximum eigenvalue is 6.099 and the CI is 0.019. The CR value is less than 0.1, which indicates a reasonable level of consistency in the pair-wise comparison that is sufficient to accept the factor weights. The weight corresponding to stream density (STE) is the highest (0.380), whereas soil (SOIL) provides the lowest number (0.035) (Tables 6.3 and 6.4).

Based on the weight of flooding hazard factors, the flooding hazard zone index (FHZI) values were calculated as:

$$\text{FHZI} = (0.248\,\text{ELE}) + (0.150\,\text{SLP}) + (0.380\,\text{STE}) + (0.037\,\text{ROC}) \\ + (0.035\,\text{SOIL}) + (0.150\,\text{MRE})$$

where ELE is the elevation factor, SLP is the slope factor, STE is the stream density factor, ROC is the rock factor, SOIL is the soil factor, and MRE is the maximum precipitation factor.

Table 6.3 AHP pair-wise comparison matrix for flooding hazard

Factors	ELE	SLP	STE	ROC	SOIL	MAXPRE	Weights
ELE	1	2	1/2	6	7	2	0.248
SLP		1	1/3	5	5	1	0.150
STE			1	7	8	3	0.380
ROC				1	1	1/5	0.037
SOIL					1	1/5	0.035
MRE						1	0.150

Maximum eigenvalue = 6.099; CI = 0.019

Table 6.4 AHP weight and class rating of flooding hazard

Factors	Weight	Class	Rating
Elevation (ELE) *[meter]*	0.248	>2000	1
		>1500–2000	2
		> 1000–1500	3
		500–1000	4
		<500	5
Slope (SLP) *[degree]*	0.150	>30	1
		>25–30	2
		>20–25	3
		10–20	4
		< 10	5
Stream density (STE) *[kilometer/sq. kilometer]*	0.380	10	1
		8	2
		4	3
		6	4
		2	5
Rock type (ROC)	0.037	Granite, Limestone	1
		Crystalic igneous, flysch	2
		alluvial	5
Soil type (SOIL)	0.035	Alisols, lixisols, ferrasols,	1
		fluvisols	2
		Water body	5
Maximum precipitation (MRE) *[mm]*	0.150	1800	1
		2000	2
		2200	3
		2400	4
		2800	5

The FHZI values are characterized by minimum (1.0), maximum (4.0), mean (2.268), and standard deviation (1.019). The FHZI is the relative hazard of a flood: the higher the value of this index, the higher hazard the area faces being flooded. These FHZI values are grouped in four classes on zones in the flooding hazard zone map: very high hazard, high hazard, medium hazard, and low hazard. Figure 6.6 shows that the Cat Thinh, Thuong Bang La, Son Thinh, Dong Khe, Thach Luong,

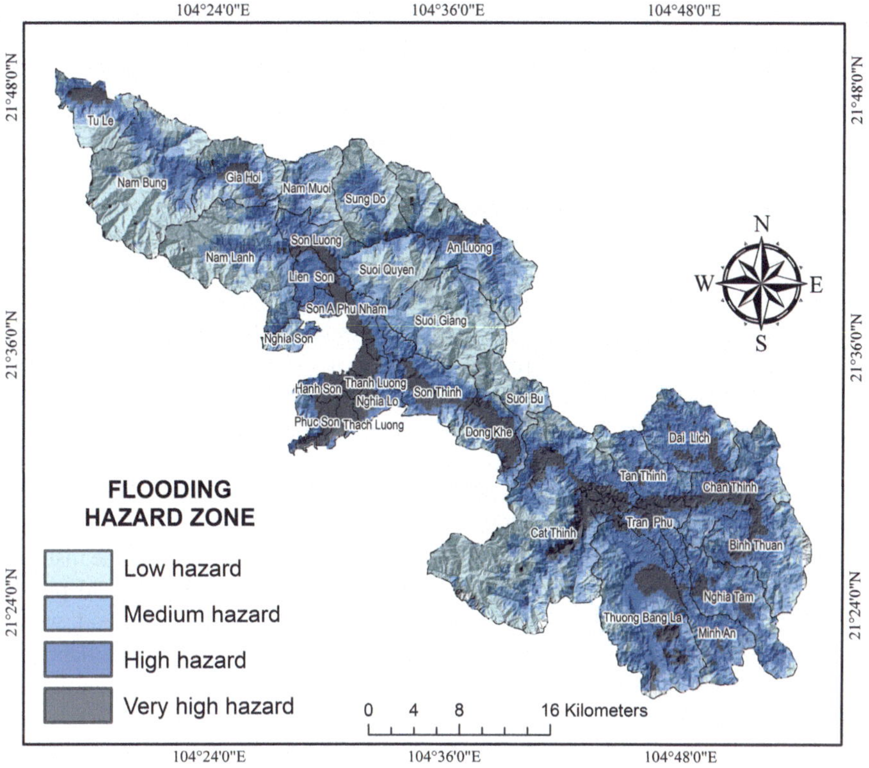

Fig. 6.6 Flood hazard zones on the Van Chan mountain. (Yen Bai Province, Vietnam)

Phuc Son, Thanh Luong, Phu Nham, and Gia Hoi communes are located in the most dangerous zone (very high hazard) for floods. Other communes are believed in safer areas (low landslide hazard), Suoi Quyen, Suoi Giang, and Suoi Bu.

6.2.2 *Gross Cost of Natural Hazards*

Storms, floods, and droughts became more frequent and intense in the Vietnamese upland in recent years. Extreme weather caused damage to the livelihoods and agriculture and affected the income of local people. Agriculture which plays a significant role in employment and poverty reduction is most affected in remote mountains. Floods and landslides damaged agriculture (yield of crops) and infrastructure (houses) and affected the local ways of living. In particular, damage corrupted between 1999 and 2016 in the Van Chan mountain is documented in the official records of the local authorities.

In 1999, most communes experienced floods after cyclones, which were accompanied by storm and heavy rains. Statistics recorded 1 death, 12 houses swept away, and 39 households evacuated from the floodplain. 10.8 hectares of upland fields experienced landslides. The storm damaged nine small and medium irrigations, 3 kilometers of canals, which irrigated over 100 hectares of wet rice paddies. The national road No. 32 to Mu Cang Chai (Yen Bai) and Than Uyen (Lai Chau) was damaged and washed away over 30 meters. An estimation of the gross cost of these damages amounted to 300,000 $US (VCG 2000).

In 2001, flash floods occurred on June 30 and August 13. These damaged 506 hectares of rice and flooded 261 hectares of crops (VCG 2002).

In 2004, long-lasting cold periods followed by heat waves affected livestock and plants. The water level in the rivers decreased by 70% and caused water shortage for irrigation. In March and April, four cyclones with hail affected communes in Van Chan. This caused damage to people and their properties. Four heavy rains caused 4 deaths and 8 collapsed houses, removed the roof of 335 houses and 10 classrooms, and broke 500 meters of aluminum cable. 15 hectares of rice and 3 hectares of crops were flooded, of which 5 hectares of rice were completed lost. Eight public constructions were damaged, causing traffic problems. An unknown number of livestock animals and poultry disappeared in the floods. The gross cost of damages was estimated at 550,000 $US (VCG 2005).

On September 18–19, 2005, heavy rains and flash floods occurred in the district. Forty-five people were killed; most of them are Kinh inhabited in fluvial plains (Fig. 6.7a). Three hundred nineteen houses collapsed as a result of the flood, while 628 were severely damaged, and 196 other houses experienced minor damage. Low-voltage and high-voltage towers, a 14,200 meter transmission line, and one transformer station were damaged resulting in electricity cuts. Eighty irrigation infrastructure assets and 23,689 meters of canals and dikes were permanently destroyed. Over 600 meters of dams were destroyed. Landslides ruined more than 1.6 kilometers of road. About 200,000 m^3 of rocks and stones blocked the road which obstructed the whole traffic system in the district. Nine concrete bridges were destroyed, and 26 suspension bridges and 99 temporary bridges were out of order. Among them, the Ngoi Thia bridge was destroyed completely by flood, and about 30 meters of road at the entrance of the Van Chan high school vanished. Mudslides blocked more than 20 sites of the national roads No. 32 and No. 37. The Ba Khe to Van Chan road was damaged over 10 meters which caused major traffic jams. Agriculture lost 678.7 hectares of rice, of which 415.8 hectares was severely damaged and 262.9 hectares permanently gone. 494.1 hectares of paddy fields faced lower yields, and 751.7 hectares of crops were damaged. Among these figures, of which, 268.1 hectares of maize and 438.6 hectares of vegetables. The floods also removed plantations, fruit trees, fishponds, livestock, and poultry. The gross cost of damages was 13.8 million $US (VCG 2006).

In 2006, a storm-induced heavy rain caused floods along the Ngoi Thia and Ngoi Nhi streams. Thirty-one houses were destroyed. Over 50 households were affected by landslides. Seventy-seven households were evacuated from the landslide areas, away from the mud and the rocks. 3.16 hectares of fields were severely flooded. 100 hectares of swamps were flooded, scattered over the communes. 29 hectares of fruit trees and hundreds of hectares of forest were cleared by hail and cyclones, which resulted in perennial crop damage. 2,000 meters of irrigation and drainage infrastructure were broken down, while 8 large and medium irrigation constructions

Fig. 6.7 Damaged site (**a**) and resettlement site (**b**) of a Kinh population in Son Thinh commune (Van Chan district) after severe flash floods occurred on September 2005. (**a**) Flash floods swept all houses and crops and killed Kinh people on fluvial plain on September 2005. (**b**) A resettlement village of Kinh was established in valley plain in 2006. (Source: An Thinh Nguyen 2015)

were damaged. Two water supply systems in the Son Thinh and Thach Luong communes were damaged. Landslides affected 2,500 meters of national, provincial roads, and rural roads. Three main state facilities and over 20 small bridges were damaged. The post office of the Nam Ben was covered by 1,000 m^3 of slurry. The gross cost of damages was 92,400 $US (VCG 2007).

In April and May 2007, two strong cyclones with hail passed by the Nghia Tam, Chan Thinh, Gia Hoi, and Nam Muoi communes. The October 5 storm was accompanied by heavy rains, which affected the infrastructure of the Ngoi Thia commune. Here 673 houses were damaged, 4 hectares of perennial crops were lost, 100 hectares of fields were flooded, and most of them were only recently planted with maize. The floods swept away three suspension bridges (including the 55 meters long bridge in Thach Luong and some sewers crossing the road). Landslides deposited about 900 m^3 of soil and stones on the access road to the village. The October floods damaged 19 small and medium irrigation dams, broke 450 meters of canals, and damaged 27 temporary dams constructed by the local farmers, which irrigated 350 hectares of rice paddies yielding two crops a year. The gross cost of the damages was estimated at 114,400 $US (VCG 2008).

In 2008, Van Chan was hit by three storms and cyclones with heavy rain and hail. This caused not only deaths but also considerable losses of agriculture and forestry. The October cyclone was accompanied by heavy rain and affected the Chan Thinh, Dai Lich, and Tan Thinh communes; this caused dead and missing persons. Local property damage included three collapsed houses and two damaged classrooms. The roof of over 100 houses was destroyed. Farmers lost 6.6 hectares of maize, 3.2 hectares of upland rice, and 4,500 square meters of fields, of which 206 hectares of paddy rice, 26 hectares of corn and, 2 hectares of acacia forest. Landslides blocked the national road No. 32. The tunnel connecting Son Luong with Sung Do in Nam Muoi was completely destroyed. The gross cost of the damages was about 110,000 $US (VCG 2009).

The rainy season of 2009 brought one to two heavy rains every month. At least three of them demand special attention. The total precipitation was 151 mm, which caused floods over a large area. Few pressure depressions during 3–4 July induced moderate heavy rain. Storm during 19–20 July brought moderate rain but was followed by heavy rain that caused damage to the local properties. Seven houses were seriously damaged and 137 houses were partially damaged. 220 hectares of crops and rice were affected. The gross cost of the damages amounted to 8,800 $US (VCG 2010).

Also 2010 brought one or two heavy storms every month during the storm and flood season: A cyclone and heavy rain affected the Son Thinh, Dong Khe, Nam Lanh, Nam Muoi, Suoi Giang, and An Luong communes during the period April 22–May 25. This caused the completed collapse of three houses. Nineteen assets of irrigation infrastructure were swept away. 200 hectares of crops and six buffalos were lost. The gross cost of the damage was about 44,000 $US (VCG 2011).

In 2011, the district was affected by three storms with heavy rain and local cyclones. Four cyclones with heavy rains passed. The storm of June 30–July 7 caused torrential rains in Cat Thinh, Chan Thinh, Nghia Son, and An Luong communes. The storm of August brought heavy rain and cyclones in the Phuc Son, Nghia Tam, and Binh Thuan communes. Cyclones during 9–17 September caused heavy rains in the Nam Lanh and Tu Le communes. The storm in October brought heavy rain on a large scale in the Nam Muoi, Sung Do, and Tu Le communes. In total the 2011 losses amounted to one death, two collapsed houses, and six destroyed houses. 10.4 hectares of fields were flooded. The irrigation infrastructure was severely damaged in the Nghia Tam, Cat Thinh, and An Luong communes. Many roads and bridges connecting villages and communes were destroyed by heavy rains and landslides. The gross cost of the damage was about 616,000 $US (VCG 2012).

In 2012, heavy rain and cyclones caused five deaths and the loss of 186 hectares of rice and corn fields and 2.2 ha of crops. Over 1,400 houses were affected, among which the roof of 1,128 building and 24 houses completely collapsed. The most damaged communes were Cat Thinh and Thuong Bang La (VCG 2013).

In 2013, three cyclones caused damage to people, houses, properties, and crops. This affected the Phuc Son, Nam Muoi, and Suoi Quyen communes. As a result 13 houses collapsed, the roof of 50 houses was demolished, 138.5 hectares of rice land were flooded, 125.3 hectares of crops damaged, and 1.79 hectares of aquaculture ponds flooded. The gross cost of the damage was established at 110,000 $US (VCG 2014).

In 2014, the district was affected by seven moderate to heavy rains and cyclones. The first cyclone happened on April 4 and caused moderate-and-heavy rain, which affected the whole Gia Hoi commune. By the end of April, storms and a cyclone damaged the Thuong Bang La, Dai Lich, Chan Thinh, and Nghia Tam communes. On May 18, cyclone-bound damage affected the whole Van Chan district. In August, rain and cyclones damaged assets in the An Luong, Phu Nham, and Hanh Son communes. During September 20–23, heavy rains affected seriously Phu Nham. Total losses in 2014 included 2 people injured and 7 houses collapsed, 244 houses of which the roof was damaged, while another 1,673 houses damaged. Nearly 700 hectares of crops were lost. Twenty-three bridges swept away, among which 19 existing ones next to two small suspensions and two cement bridges. The cyclones also damaged 7 dams and 70 meters of dykes. The gross cost of damages was estimated at 200,000 $US (VCG 2015).

In 2015, five extreme cold periods occurred, which lasted more than 3 days each. 17 to 23 December was a period of extreme cold with frost in high mountains of Suoi Giang, Nam Bung, Gia Hoi, Nam Muoi, and Sung Do. Minimum temperatures ranged between 6.4 and 7.2^{0}C. In June 2015, Van Chan faced rains storms and cyclones, which caused flash floods and floods all over the district. In August 2015, heavy rain and cyclones damaged vast areas (Thach Luong, Son A, Nam Lanh, Thanh Luong, Suoi Bu, Phuc Son, Nam Bung, Son Luong, Phu Nham, and An Luong) and affected 224 houses. 29 hectares of rice were flooded and lost next to 6 hectares of crops. 39 hectares of fishponds were damaged in Chan Thinh, Tan Thinh, and Dai Lich communes. The gross cost of damages was about 22,000 $US (VCG 2016a) (Fig. 6.8).

Table 6.5 summarizes the extreme weather events and their impacts. Overall, table shows that Van Chan is seriously damaged by multi-hazards during the rainy season. Storms and heavy rains triggered cyclones, floods, flash floods, and landslides, which caused most of the damage, affecting all aspects of local livelihood. Life in local communities appears as safe during the period between the November and March: no heavy rain and related natural hazards were found during this time. However, extreme colds which had negative impacts on agriculture appear from December to February. Only in November 2008 and 2014, extreme colds happened.

Fig. 6.8 Flash floods damaged crop fields and electricity and transport infrastructures along provincial road in the Van Chan district (Source: An Thinh Nguyen 2017)

Drought occurred during summer and damaged crops mainly paddy rice, corn, and fruit trees. Cat Thinh, Thuong Bang La, Dong Khe, Son Thinh, An Luong, Nam Lanh, Nam Bung, Tu Le, Thach Luong and Phu Nham are communes most vulnerable to natural hazards.

6.3 Conclusion

A review of extreme weather and its effects shows that the Van Chan district is yearly affected by devastating weather during the rainy season which last from July until October. During wintertime the region experiences regularly cold periods and frost. Drought occurs every 3 years on average from April to June. These heavy weather conditions affect severely agricultural yields, primarily rice, maize, vegetables and fruits, infrastructure and houses, and the life of these inhabitants of the mountains. The costs to repair the direct damage regularly exceed the focal budgets.

An analysis of the factors influencing floods and landslides show a differential risk for a range of landscapes in the Van Chan mountain. Overall the data show that in Vietnam extreme weather not only affects the more intensively studied coastal zones and estuary plains but also the household income and way of life in this remote and poor mountainous district.

Table 6.5 Official statistics of damage by natural hazards between 1999 and 2016 in the Van Chan district (Yen Bai Province, Vietnam)

Year	Most serious natural hazards	Natural hazard period												Most affected areas	Estimated total cost
		Jan	Feb	Mar	Apr	May	Jun	Jul	Aug	Sep	Oct	Nov	Dec		
1999	Heavy rain													Cat Thinh, Tu Le	300,000 $US
	Flash flood														
	Landslide														
	Drought														
2004	Heavy rain													Chan Thinh, Cat Thinh, Thuong Bang La, Suoi Giang	550,000 $US
	Flash flood														
	Cyclone														
	Drought														
	Extreme cold														
2005	Storm													Phu Nham, Son Thinh, Chan Thinh, Cat Thinh, Suoi Giang	13.8 million $US
	Flood														
	Flash flood														
	Landslide														
2006	Storm													Nghia Tam, Minh An, Phu Nham, Son Thinh	92,400 $US
	Flood														
	Landslide														
2007	Cyclone													Chan Thinh, Son Luong, Gia Hoi	114,400 $US
	Flash flood														
2008	Storm													Chan Thinh, Dai Lich, Tan Thinh, Binh Thuan	110,000 $US
	Cyclone														
	Heavy rain														
	Flood														
	Extreme cold														

(continued)

Table 6.5 (continued)

Year	Most serious natural hazards	Natural hazard period												Most affected areas	Estimated total cost
		Jan	Feb	Mar	Apr	May	Jun	Jul	Aug	Sep	Oct	Nov	Dec		
2009	Heavy rain													Dong Khe, Ngoi Thia	8,800 $US
	Cyclone														
2010	Heavy rain													Son Thinh, Dong Khe, Nam Lanh	44,000 $US
	Cyclone														
2011	Storm													Binh Thuan, An Luong, Nghia Tam, Cat Thinh, Tu Le	616,000 $US
	Cyclone														
	Heavy rain														
	Flash flood														
	Landslide														
	Extreme cold														
2012	Storm													Cat Thinh, Thuong Bang La	N/A
	Heavy rain														
	Cyclone														
2013	Storm													Phuc Son, Nam Muoi, Suoi Quyen	110,000 $US
	Heavy rain														
	Flood														
	Cyclone														
2014	Heavy rain													Phu Nham, Nghia Lo, Thuong Bang La, Chan Thinh	200,000 $US
	Cyclone														
	Extreme cold														
2015	Heavy rain													Suoi Giang, Nam Bung, Thuong Bang La, Chan Thinh	22,000 $US
	Cyclone														
	Flood														
	Extreme cold														
	Drought														

Source: Official statistic data provided by Van Chan authorities, 2000–2016

Chapter 7
Comparing Local and Immigrant Household Preparedness for Natural Hazards in the Van Chan Mountain

Abstract Vulnerable groups are affected more by extreme weather as compared to the population on average. In the Vietnamese mountains, the ethnic minorities, as the Tay, Thai, Hmong, and Kinh, are definitely among the most vulnerable groups in the country: they are poor and live on minimal resources stemming from traditional agriculture (tea, rice, and vegetables). The human ecological characteristics of these groups are poorly described in the literature.

This chapter aims at identifying which natural hazards these populations perceive as threatening and how to adapt to them. A major aspect of the discussion relates to the changing LULC in the area.

The chapter takes off with a description of the LULC changes on the Van Chan mountain. Villages and agriculture occupy gradually but deliberately more land, while deforestation goes on. The four main ethnic minorities (Tay, Thai, Hmong, and Kinh) living in this area are described in their human ecological context. A two-round Delphi-inspired analysis in which representatives of these four ethnicities took part allows identifying the most important natural calamities as floods and landslides, together with the prioritization of their impacts. The research equally allowed inventorying the adaptation actions already undertaken by the local inhabitants.

Keywords Preparedness · Village-scapes · Delphi survey · Local experience and responses · Statement statistics · Van Chan mountain

7.1 Village-Scapes of Local and Immigrant Communities

Van Chan ranks among the most remote and poorest districts in the Yen Bai province. Among the 31 communes of the Van Chan district, 16 are part of the national program for hunger and poverty alleviation as communes facing particular difficulties (VCG 2016b). The district has a population of approximately 150,000 people spread over more than 35,000 households. Among the 18 ethnic groups found in the area, 8 have a population above 1000 inhabitants: Kinh (34.3%), Thai (23.4%), Tay (17.1%), Dzao (9%), Hmong (7.1%), Muong (2.1%), Dzay (1.3%), and Kho-mu (0.7%) (VCG 2012) (Fig. 7.1). The most numerous population of Kinh immigrated to Van Chan mountain mainly from the Red River Delta since the 1960s. The other ethnicities are indigenous.

A. T. Nguyen, L. Hens, *Human Ecology of Climate Change Hazards in Vietnam*, Springer Climate, https://doi.org/10.1007/978-3-319-94917-8_7

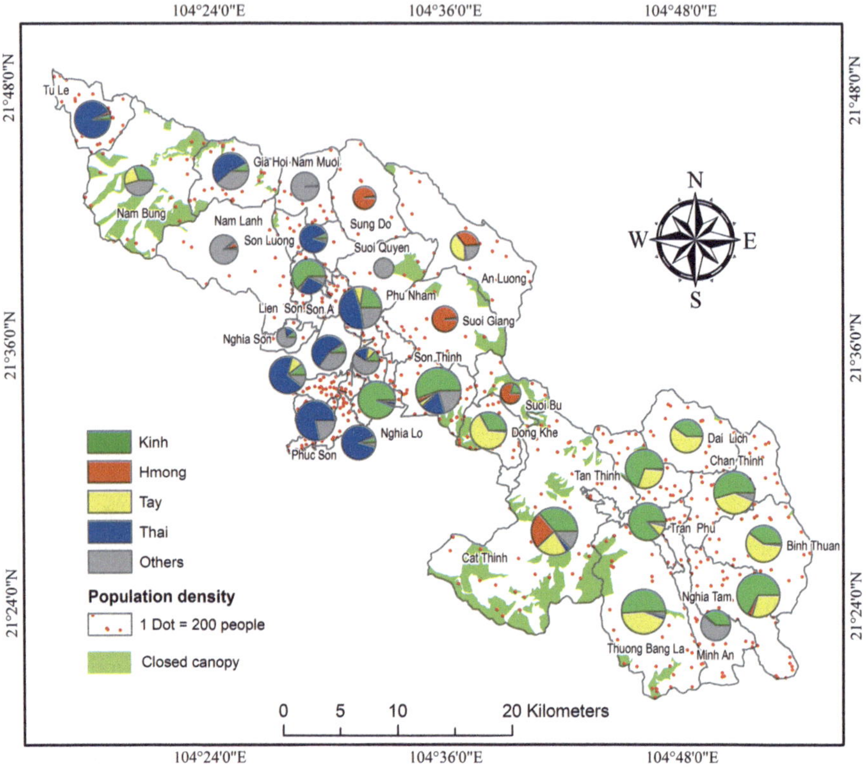

Fig. 7.1 Distribution of ethnic groups in the Van Chan mountain. (Yen Bai province, Vietnam)

Table 7.1 LULC changes during 1993–2013 in the Van Chan mountain (hectares)

LULC	1993	2003	2013	1993–2003	2003–2013	1993–2013
Paddy	55,936	50,481	60,528	−5455	10,047	4592
Forest	63,860	63,225	51,798	−635	−11,427	−12,062
Built-up	4166	9959	11,364	5794	1404	7198
Water	46	342	318	296	−24	272
Total	*124,008*	*124,008*	*124,008*			

Table 7.1 and Figs. 7.2, 7.3, and 7.4 show the LULC changes in area in the Van Chan district during the most recent 20 years. The rice paddies cover the largest area (60,528 hectares in 2013), next to forests (51,798 hectares), while build zones (11,364 hectares) and water bodies (318 hectares) cover the smallest part of land. Paddy land covered 10,047 hectares in the 2003–2013 period, increasing 4592 hectares during 1993–2013, built-up land (5794 hectares during 1993–2003, 1404 hectares during 2003–2013, and 7198 hectares during 1993–2013), and water bodies (272 hectares during 1993–2013). The forest surface decreased (635 hectares during 1993–2013, 11,427 hectares during 2003–2013, and 12,062 hectares during 1993–2013). Over two decades, built-up and paddies increased most (7198 hectares and 4592 hectares, respectively), whereas forest land cover decreased (12,062 hectares).

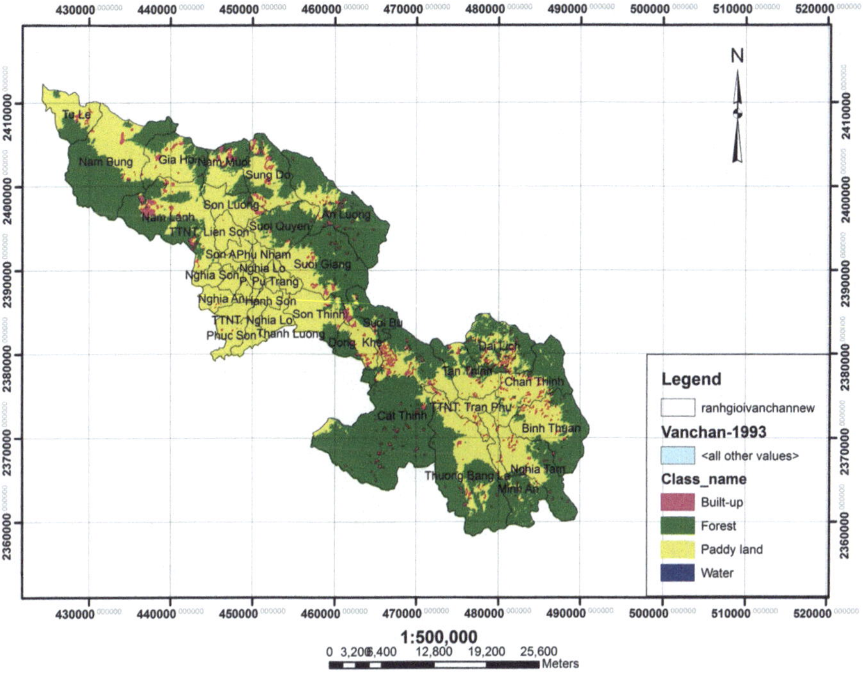

Fig. 7.2 Land use land cover map of the Van Chan mountain. (Yen Bai province, Vietnam) in 1993

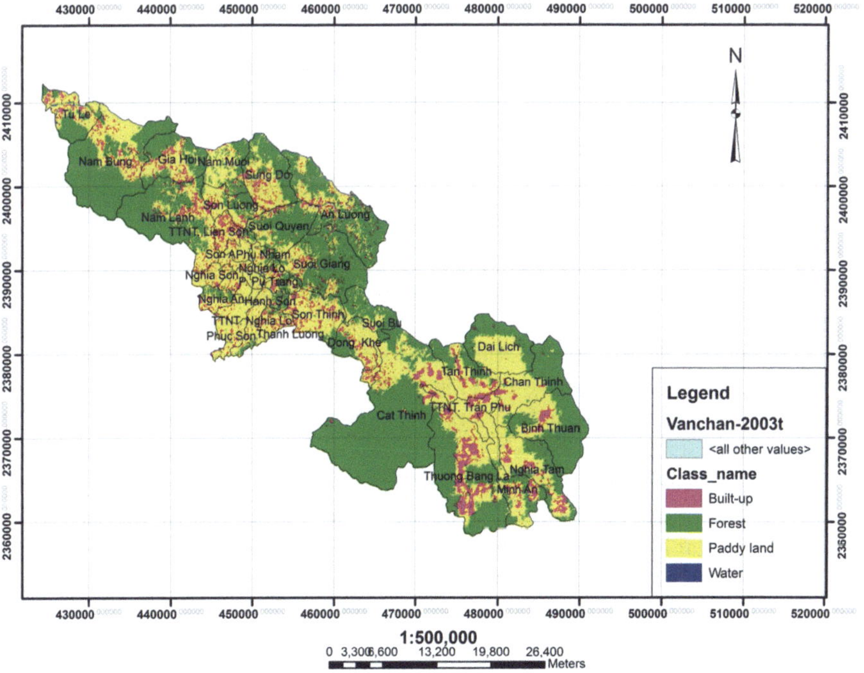

Fig. 7.3 Land use land cover map of the Van Chan mountain. (Yen Bai province, Vietnam) in 2003

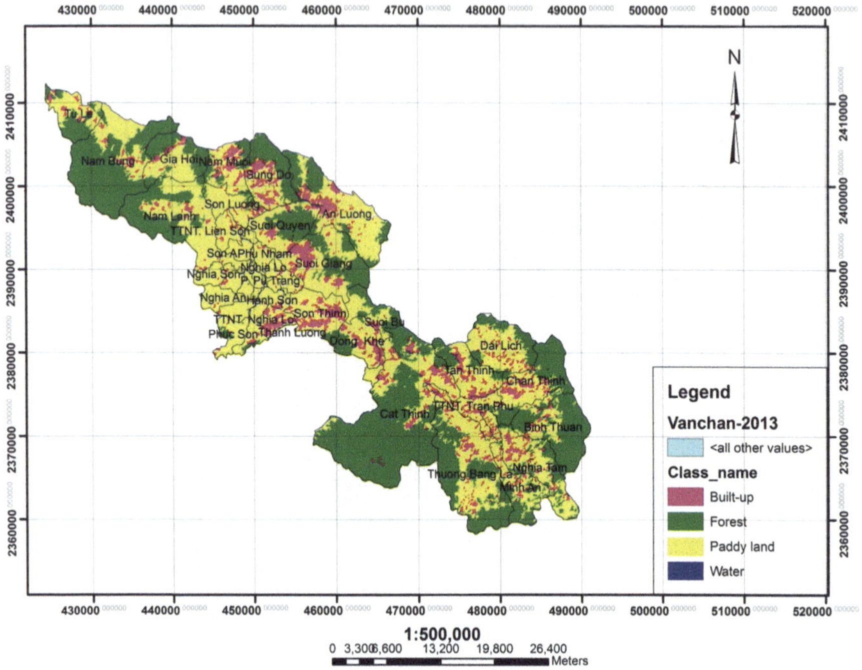

Fig. 7.4 Land use land cover map of the Van Chan mountain. (Yen Bai province, Vietnam) in 2013

This shows the increasing urbanization and agricultural expansion in contrast with the deforestation in Van Chan mountain during recent years.

The following village-scapes in the Van Chan mountain were used as case studies:

(a) *Village-scapes of Thai in the Son Thinh commune*

The Son Thinh commune is part of the Muong Lo valley which covers 3151 hectares. The valley is pan-shaped and has a diversified topography, entailing high mountains in the north and a hilly zone in the east. The case study terrain entails three zones: a central zone (of flat delta), a high mountainous zone in the northeast (at an altitude of 500–600 meters), and a hilly zone (Fig. 7.5). The terrain slopes from northeast to southwest Son Thinh is the district capital and the homeland of the Vietnamese Thai population. The Thai make up 17% of the population in the commune. They traditionally live and farm near the rivers and springs. Water resources are crucial for their agriculture. Indigenous knowledge allows people growing rice on irrigated paddies and vegetables on the slopes of the hills and mountains which provides a unique landscape. Agriculture and forestry are the two main sources of income (76% of the total income). On average the individual income is about 355 $US per year (STG 2016). The district land use planners consider the valley (including part of the Son Thinh commune) as protected land, which encourages the development of forestry.

Fig. 7.5 Paddy rice and crops are cultivated by the Thai along the river and on the slopes of the Son Thinh commune (Van Chan mountain). (Source: Nguyen An Thinh 2016)

The national road No. 32 crosses the commune, which contributes to the local socio-economic development and facilitates trade between the commune and the neighboring areas in the northern mountain and the Red River Delta.

(b) *Village-scapes of Tay in the Thuong Bang La commune*

The Thuong Bang La commune covers 9243 hectares and is located on one side of the Lung Lo hill. The commune is famous for the Tay's culture. Tay accounts for nearly 44% of the population in the village (TBLG 2016). Traditional villages of the Tay are at the foot of mountains and on the edge of hills (Fig. 7.6). Tay farmers cultivate rice, maize, and sweet potato. Because most of fields are in the plains, they are more vulnerable to heavy rains, floods, landslides, and cyclones.

Fig. 7.6 Paddy rice is cultivated by the Tay on the fertile plains of Thuong Bang La commune (Van Chan mountain). (Source: Nguyen An Thinh 2016)

(c) *Village-scapes of Hmong in the Suoi Giang commune*

Suoi Giang is a commune high in the mountains of the Van Chan district. The village is 12 kilometers away from the district center. Suoi Giang is in the mountains, 1371 meters above the sea level, which results in comparatively low temperatures as compared to other areas in the province (Fig. 7.7). The difference between the average temperature of Suoi Giang and that of Van Chan is approximately 9 °C. Hmong inhabitants in the commune account for 98.2%; the rest are Kinh, Dzao, and Tay (SGG 2016).

Suoi Giang is famous for a type of green tea called *San Tuyet Suoi Giang tea* ("San Tuyet" means "mountainous snow"). The tea leaves are collected 7 years after the trees were planted. The tea trees are grown without chemical fertilizers and herbicides, which declines the quality of tea production. Growing, maintaining, and collecting tea trees are the main livelihood activities of the Hmong in

Fig. 7.7 The biological "San Tuyet Suoi Giang tea" is cultivated by the Hmong on the slopes of Suoi Giang commune (Van Chan mountain). (Source: Nguyen An Thinh 2016)

Suoi Giang. Tea leaves are harvested four times per year, in contrast with two harvests per month for tea grown in the midlands. The price of San Tuyet Suoi Giang tea is about eight times that of tea grown in the midlands. Moreover, the price of tea harvested in winter is about 11 times that produced from tea trees grown in the midlands. As a result, the San Tuyet Suoi Giang tea is relatively profitable for the Hmong farmers.

Besides tea, food crops (e.g., corns, rice and cassavas) are grown. These crops serve mainly the self-sufficiency or the small-scale local trade. Agroforestry includes cinnamon, acacia, and eucalyptus. Tourism also reached the Mang Cang cultural village where Hmong are at home.

(d) *Village-scapes of Kinh in the Tran Phu town*

The Tran Phu town is southwest of the Thuong Bang La commune, from which it is separated by a hill. The Tran Phu commune is spread over 1892 hectares.

Fig. 7.8 Mixed cultivation systems (forest, tea, orange, paddy rice, crops, and small fishponds) of the Kinh on sloping lands of Tran Phu town (Van Chan mountain). (Source: Nguyen An Thinh 2016)

A majority of 86% of the population is Kinh, which corresponds with 9.46% of the Kinh inhabitants of the Van Chan district (TPG 2016). The commune was founded as a large state-managed farm established since 1967. Thousands of Kinh from the Red River Delta immigrated to this area, exploited forests, and built farms. Kinh farmers in Tran Phu have grown green tea, for which the commune is famous. Unfortunately, the tea generated a modest income. Since the 1990s, Kinh farmers have cultivated oranges, replacing the tea plants on the lower slopes. The orange provides high yields and an attractive income. This improved the quality of life of the farmers and their families. Mixed cultivation systems are expanded: farmers grow orange trees on the modestly inclining slope land and tea trees on the midland, while forests remain on the steep slopes (Fig. 7.8). Today green tea trees are still the dominant crop in Tran Phu, whereas orange trees provide more income.

7.2 Delphi Round 1

Both local and immigrant communities who inhabit in the most damaged areas were selected for the survey. Four groups of indigenous panel respondents were selected for the stakeholder Delphi survey: Thai (in Son Thinh commune), Tay (Thuong Bang La commune), Hmong (Suoi Giang commune), and Kinh (Tran Phu town). Impacts of natural hazards and local responses before, during, and after hazards were identified. The Delphi process uses six questions in the first round and three questions in the second round. These questions were listed according to impacts and responses. The four questions on the impacts concern list of natural hazards, the most severe local hazards, drivers of natural hazards, and identification of natural hazard hotspots. Two questions on the local responses include local experience and list of response activities by local people before, during, and after heavy rains.

7.2.1 Impacts

As shown in Table 7.2, four questions deal with the impacts of climate change hazards and extreme weather events on agriculture.

Natural hazards before heavy rain (Q_1.1) Cyclones caused the greatest damage for Tay (42/50) followed by flash floods (10/50) to Hmong.

Natural hazards during heavy rain (Q_1.1) Cyclones, floods, and landslides damaged Thai (40/50, 34/50, and 30/50, respectively) and Kinh (34/50, 43/50, and 39/50), while Tay (50/50) agree that floods cause most damage.

Natural hazards after heavy rain (Q_1.1) Landslides were mentioned by all four ethnic groups as causing the most impacting damage to the local communities (20/50, 31/50, 25/50, and 7/50).

Among the natural hazards, floods are most severe as acknowledged by the Thai (34/40), Tay (45/50), and Kinh (39/50) (Q_1.2).

Several drivers of natural hazards are reported by the locals (Q_1.3). Heavy rain is most frequently mentioned especially by the Tay (47/50). Both slopes (35/50) and lowlands (26/50) are vulnerable to natural hazards according to local people. The damage by the natural hazards intensified as a result of the low quality of drainage system, as stated by the Thai (14/50) and the Tay (8/50). Temporary houses of the poor in the uplands are one of the principal drivers of the severe impacts of natural hazards.

The locations of natural hazards most frequently reported by the Thai include Suoi Son (16/50), Ban Hoc (10/50), and Van Thi (4/50). Tay reported hotspots as Nga-ba-nong-truong (6/50) and Lung Lo (17/50) in Thuong Bang La. Three hotspots in Suoi Giang were reported by Hmong: Suoi Lop (28/50), Giang Cao (20/50), and Tap Plang (13/50). Suoi Lao is the only one hotspot reported by the Kinh of Tran Phu.

Table 7.2 Questions on impacts (I) of natural hazards and replies by the respondents in round 1

Code	Statements	Most frequently selected alternative (number of respondents)				Symbol of the round 2 statement (S-code)
		Thai (Son Thinh)	Tay (Thuong Bang La)	Hmong (Suoi Giang)	Kinh (Tran Phu)	
Q_1.1	Natural hazards before heavy rain	Cyclones (16/50)	Cyclones (42/50)	Flash floods (10/50) Cyclones (10/50)	Cyclones (7/50)	Natural hazards cause the greatest damage for locals (S_1)
	Natural hazards during heavy rain	Cyclones (40/50) Floods (34/50) Landslide (30/50)	Floods (50/50) Cyclones (33/50)	Cyclones (28/50) Landslides (25/50)	Floods (43/50) Landslides (39/50) Cyclones (34/50)	
	Natural hazards after heavy rain	Floods (24/50) Landslides (20/50)	Landslides (31/50)	Landslides (25/50)	Landslides (7/50) Cyclones (7/50)	
Q_1.2	Most severe local hazards	Floods (34/50)	Floods (45/50)	Landslides (33/50)	Floods (39/50)	
Q_1.3	Drivers of natural hazards	Heavy rain (28/50) Lowland (valley), high density of river network (26/50) Infrastructure: low quality of the drainage system (14/50) Nonpermanent house (24/50)	Heavy rain (47/50) Infrastructure: low-quality drainage system (8/50) Destroy headwater forest (11/50) Nonpermanent houses (6/50)	Sloping land, heavy rain (35/50) Transport construction (33/50) Temporary housing construction (20/50)	Heavy rain (36/50) Inefficient drainage system (7/20).	The principal drivers of natural hazards (S_2)
Q_1.4	Natural hazard hotspots	Suoi Son (16/50) Ban Hoc (10/50) Van Thi (4/50)	Nga-ba-nong-truong (6/50) Lung Lo (17/50)	Suoi lop (28/50) Giang Cao (20/50) Tap Plang (13/50)	Suoi Lao (34/50)	

7.2.2 *Local Experience and Responses*

Local experience (Q_2.1) Table 7.3 shows the answers to the questions on local responses (R) to natural hazards and their selection incidence. Asked about the local experience (Q_2.1) with natural hazards, the Hmong in Suoi Giang pointed to mitigation measures they established: gathering cattle to barns (10/50), covering the barns (10/50), planting forests to prevent erosion and landslides (7/10), building low-ceiling houses leaning on the slope to reduce the impacts of wind

Table 7.3 Questions on local responses (R) to natural hazards and reply by the respondents in round 1

Code	Statements	Most frequently selected alternative (number of respondents)				Symbol of the round 2 statement (S-code)
		Thai (Son Thinh)	Tay (Thuong Bang La)	Hmong (Suoi Giang)	Kinh (Tran Phu)	
Q_2.1	Local experience	Cover barns (6/50) Alarms for evacuation (2/50) Build low-ceiling houses (2/50)	Observing changes in the sky and plants (3/50) Harvest before storm season (3/50) Gather cattle to barns (8/50) Cut tree branches around houses (3/50) Grow trees to prevent landslides (3/50) Reduce deforestation(6/50) Escape to tea growing hills (6/50)	Support from local authorities (10/50) Gather cattle to barns (10/50) Grow trees to prevent erosion, landslides (7/50) Build low-ceiling houses; lean on slopes to reduce impacts of wind (5/50) Share information with each other (3/50) Cover barns (10/50) Reinforce buildings with stones (2/50) Avoid crossing big streams (3/50) Hit the fireplace (8/50) Grow plants outside the natural hazards (7/50)	Avoid going out (11/50) Afforestation (11/50) Evacuate from hazard hotspots (7/50) Weather forecasts by noticed changes (5/50)	The local experiences to natural hazard mitigation (S_3)
Q_2.2	Response befcre heavy rain	Hazard warnings information from the local authorities (42/50) Information by television and radio (28/50) Prepare foods in place (22/50) Reinforce buildings and roofs (19/50)	Hazard warning from local government (45/50) Reinforce houses and roofs (33/50) Temporal migration from hazard hotspots (50/50)	Hazard warning information from local authorities (48/50)	Hazard warning information from local authorities (46/50)	Local response before heavy rain (S_4)

(continued)

Table 7.3 (continued)

Code	Statements	Most frequently selected alternative (number of respondents)				Symbol of the round 2 statement (S-code)
		Thai (Son Thinh)	Tay (Thuong Bang La)	Hmong (Suoi Giang)	Kinh (Tran Phu)	
	Response during heavy rain	Receive timely hazard information from the local authorities (40/50) Receive information from television and radio (38/50) Temporal migration from hazard hotspots (22/50)	Receive information from television and radio (36/50) Receive timely hazard information from local government (25/50)	Information from television and radio (45/50)	Information from television and radio (41/50)	Local response during heavy rain (S_5)
	Response after heavy rain	Hazard warning information from the local authorities (40/50) Information from television and radio (42/50) Store foods in house (26/50)	Information from television and radio (25/50) Timely hazard information from the local authorities (31/50)	Store foods in house (33/50)	Store foods in house (23/50)	Local response after heavy rain (S_6)

(5/50), sharing information with each other (3/50), fortifying with stone building (2/50), avoiding crossing big streams (3/50), growing plants during the non-hazard season (7/50), and receiving support from the local government (10/50). From experiences, Tay forecast natural hazards based on changes in the sky or the plants (3/50).

Response before heavy rain (Q_2.2) The local government adopts an important role in disseminating information about natural hazards in beforehand. Most of the respondents receive warnings from local authorities: Thai (42/50), Tay (45/50), Hmong (48/50), and Kinh (46/50) all mention this aspect. The Thai also retrieve information from television and radio (28/50). Both Thai and Tay reinforce their houses and roofs (19/50 and 33/50, respectively). Tay migrates from hazard hotspots (50/50).

Response during heavy rain (Q_2.2) Thai (38/50), Tay (36/50), Hmong (45/50), and Kinh (41/50) all retrieve updated information from television and radio.

Response after heavy rain (Q_2.2) Hazard information by the authorities and in the media after heavy rain is followed up by all four populations: Thai (40/50), Tay (31/50), Hmong (33/50), and Kinh (23/50).

7.3 Delphi Round 2

7.3.1 Statement Statistics

Delphi round 2 provides the coherence and the reliability of the Delphi results: the statements were refined using the most frequently mentioned alternatives in round 1. Table 7.4 shows the statistics of the results of each statement in round 2. Mean values indicate the degree to which local people support or prioritize the statements on a 5-point Likert scale. Standard deviations vary over a small range (below 2.0) and indicate that the variation between the responses is limited and that the information provided by the stakeholders is consistent. It also points to a consensus among panel members. The quartile value provides an indication of the variation and the dispersion of the data sets. For the different statements, the benchmark scores are almost equal or only slightly divergent. This confirms that the impacts of natural hazards, local experience, and local responses on different populations are perceived in a similar way.

- *Impacts:* Statement S_1 attracts the highest values of the mean for Thai (3.96), Tay (4.00), and Hmong (4.08) alike. Statement S_2.1 realizes the highest mean score (3.92) for the Kinh.

Table 7.4 Statistics of the responses to the statements of Delphi round 2

Statements			Mean	Rank		Standard deviation	Min.	Max.	Percentiles		
				Mean	Score				25th	50th	75th
(a) Thai (Son Thinh commune)											
Impact	Natural hazards cause the greatest damage for locals	S_1	3.96	6	1	1.39	1	5	3.00	5.00	5.00
	The major drivers of natural hazards	S_2	3.76	4	3	1.44	1	5	2.75	4.00	5.00
Local experiences	The local experiences to natural hazard mitigation	S_3	3.50	3	4	1.30	1	5	2.75	4.00	5.00
Local responses	Local response before heavy rain	S_4	2.64	1	6	1.22	1	5	2.00	2.00	4.00
	Local response during heavy rain	S_5	3.88	5	2	1.24	1	5	3.00	4.00	5.00
	Local response after heavy rain	S_6	3.38	2	5	1.18	1	5	2.75	3.00	4.00
(b) Tay (Thuong Bang La mountain)											
Impact	Natural hazards cause the greatest damage for locals	S_1	4.00	5.5	1	1.40	1	5	3.75	5.00	5.00
	The major drivers of natural hazards	S_2	3.80	4	2	1.43	1	5	2.75	4.00	5.00
Local experiences	The local experiences to natural hazard mitigation	S_3	3.56	2.5	3	1.37	1	5	2.75	4.00	5.00
Local responses	Local response before heavy rain	S_4	1.74	1	4	1.21	1	5	1.00	1.00	2.00
	Local response during heavy rain	S_5	4.00	5.5	1	1.16	1	5	3.75	4.00	5.00
	Local response after heavy rain	S_6	3.56	2.5	3	1.23	1	5	3.00	4.00	5.00
(c) Hmong (Suoi Giang mountain)											
Impact	Natural hazards cause the greatest damage for locals	S_1	4.08	6	1	1.05	1	5	4.00	4.00	5.00
	The major drivers of natural hazards	S_2	3.98	5	2	1.15	2	5	3.00	4.00	5.00
Local experiences	The local experiences to natural hazard mitigation	S_3	3.36	3	4	1.12	2	5	2.00	3.00	4.00

(continued)

Table 7.4 (continued)

Statements			Mean	Rank Mean	Rank Score	Standard deviation	Min.	Max.	Percentiles 25th	Percentiles 50th	Percentiles 75th
Local responses	Local response before heavy rain	S_4	3.66	4	3	1.06	1	5	3.00	4.00	4.00
	Local response during heavy rain	S_5	2.70	2	5	1.30	1	5	1.00	3.00	4.00
	Local response after heavy rain	S_6	2.02	1	6	1.27	1	5	1.00	1.50	3.00
(d) Kinh (Tran Phu mountain)											
Impact	Natural hazards cause the greatest damage for locals	S_1	3.68	5	2	1.27	1	5	3.00	4.00	5.00
	The major drivers of natural hazards	S_2	3.92	6	1	1.18	2	5	3.00	4.00	5.00
Local experiences	The local experiences to natural hazard mitigation	S_3	3.42	3	4	1.14	2	5	2.00	3.00	4.25
Local responses	Local response before heavy rain	S_4	3.62	4	3	1.05	1	5	3.00	4.00	4.00
	Local response during heavy rain	S_5	2.68	2	5	1.32	1	5	1.00	3.00	4.00
	Local response after heavy rain	S_6	1.82	1	6	1.06	1	4	1.00	1.00	2.25

- *Local experience:* Statement S_2 scores average (3.36–3.56). This reveals that the local experiences in hazard mitigation such as alarms for evacuation, building low-ceiling houses, and observing changes in the sky and on the indicator plants are less supported by local people.
- *Local responses:* Statement S_4 scores average to low (1.74–3.66). This means that local responses to natural hazards are considered inadequate. Statement S_5 scores a high mean both among the Thai and Tay (3.88 and 4.00, respectively), while it is low for Hmong and Kinh (2.70 and 2.68, respectively). Each ethnic group has a different priority on the statement about local responses during heavy rain. The most common responses include receiving natural hazard information from the local authorities or via the media and migrating from hazard hotspots. Most of the respondents express low support toward statement S_6 as reflected by the low score of the mean value (1.82–3.38). The responses after heavy rain also include receiving information from several sources such as the local authorities and the media.

Figure 7.9 represents the mean values of the statements. The results show that different areas have different natural hazards and drivers. There are similarities and differences in the preparation for hazard prevention among ethnic groups. Indigenous knowledge on natural hazard prevention in the communes differs among the ethnic groups. Statement S_5.2 attracts the highest mean (9.26), reflecting the highest

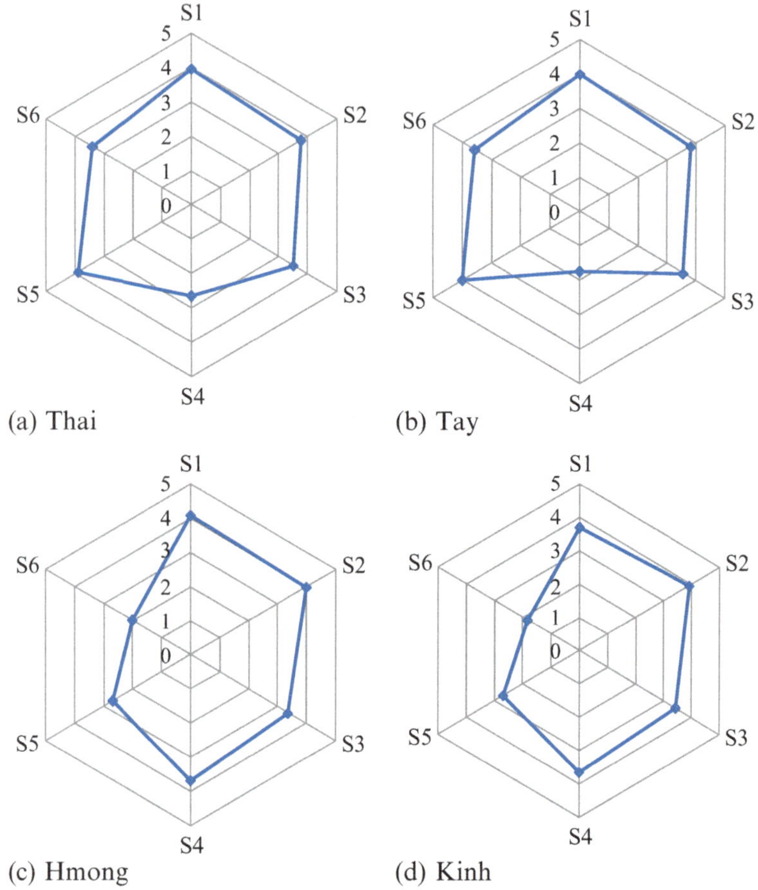

Fig. 7.9 Mean values in round 2 of the Delphi survey statements

priority for the stakeholders. The mean values of statements S_2.3, S_3.2, and S_4.3 are higher than or equal to 9. Statements with a low score include S_4.4 (5.06/10), S_3.5 (5.75/10), and S_5.4 (6.62/10).

7.3.2 *Kendall's W Test*

The responses by the panel members collected during the second Delphi round allow calculating a Kendall's W of 0.681. As shown in Table 7.5, there are "strong" to "very strong" agreement of consensus among the panel members and the "high" to "very high" confidence in ranks. Therefore, Delphi process finished at the second round. Kendall's W of the Thai is the lowest (0.603). The Tay and Kinh populations attract higher scores (0.617 and 0.772, respectively). The highest confidence is 0.772 by Hmong.

Table 7.5 Kendall's W for weights assigned during the second Delphi round

Population	N	Kendall's W	Chi-Square	P	Agreement	Confidence in ranks
Thai (Son Thinh)	50	0.603	150.676	<0.001	Strong	High
Tay (Thuong Bang La)	50	0.617	154.294	<0.001	Strong	High
Hmong (Suoi Giang)	50	0.772	193.008	<0.001	Very strong	Very high
Kinh (Tran Phu)	50	0.759	189.823	<0.001	Very strong	Very high

7.4 Conclusion

The data in this chapter reveal a differentiated picture of the hazards and the way the local ethnicities cope with them. Points of attention include:

- Land use changes on the Van Chan mountain point to an increasing urbanization and a gradually vaster area for agriculture of mainly rice and tea. The land necessary for this dynamic comes from deforestation. This process is not unique for this region. The same trends are observed in other (fast) developing parts in Vietnam, north central, northeast, central highland, and northwest areas (Khuc et al. 2018). Moreover this type of pressure on nature is not only described in Vietnam but also in other countries of Southeast Asia, in Africa, and in Latin America (Subedi et al. 2014; Dohong et al. 2017; Armenteras et al. 2017).
- Van Chan is the homeland of four studied ethnic minority groups: Tay, Thai, Hmong, and Kinh. Their income is based on agriculture, although there is a differentiation in their products. While the Thai and the Tay are mainly rice farmers on the worldwide known terraced slopes, the Hmong are famous for their (rare) bio-tea, and the Kinh derive a significant income from oranges. In combination with the cultural authenticity of the area, these features have a tourism potential in common. Developing tourism might raise the (limited) household income in this region and expand its economic basis.
- Cyclones, floods, and landslides are identified by the inhabitants as the main natural hazards. They cause impacting damage to temporary constructions and roads. This is also the case in Ky Anh and emerges as a more general phenomenon. The location of the most affected spots differentiates the villages.

The study results show that limited local experience on the protection of properties and public assets exists. More effective support depends on public policies. Actions by the local population strongly depend on the information on hazards by the authorities. Also these aspects are not unique. The demand for more and more reliable information on upcoming natural catastrophes also exists in Ky Anh. There is ample space for the local implementation of countrywide adaptation policies seriously enhancing the efficiency of how the local inhabitants protect their own and the public assets.

Part IV
Policy Implications

Chapter 8
Building Resilient Landscapes and Green Cities Along the Coast and in the Upland of Vietnam

Abstract This concluding chapter aims at integrating the data of the previous chapters. It starts from the observation that the reviewed and original data from the core chapters provides a multitude of research facts on Vietnam which leave few of any doubt on climate change associated effects such as storms, inundations, coastal erosion, and their significant impacts on the nature and the affected human populations. Two main recommendations are provided. Firstly, rethink cities and communes in a context of sustainability and green and smart organizations. Adopt an integration approach targeting not only sufficient accessible green but also, among others, carbon neutrality, zero waste, sustainable mobility, an urban economy with opportunities for well-being, and an inclusive and safe social network. Use smart solutions for these ambitious targets. Secondly, strengthen in Vietnam a provincial and communal "disaster policy" which is aligned with the comprehensive national climate change policy.

The chapter concludes by focusing on the uncertainties associated with the research results in this book, among others about the cost of the damage caused by extreme weather conditions.

Keywords Policy implications · Risk · Resilient landscape · Green city · Coastal protection · Sustainable land use · National challenges

8.1 Human Ecological Aspects of Climate Change Hazards at a Local Scale

8.1.1 Risks for Nature and Humans in Lowlands

Vietnam is prone to effects of climate changes: the country experiences increasing temperature variations, sea level rise, and an increasing variation in the onset and the end of its main seasons (MONRE 2016). Particularly, the coastal plains of Central Vietnamese lowland are cut by small, short, and steep rivers (often overflowing during the rainy season), which makes the area vulnerable by extreme weather and natural hazards, as storms and floods.

A. T. Nguyen, L. Hens, *Human Ecology of Climate Change Hazards in Vietnam*, Springer Climate, https://doi.org/10.1007/978-3-319-94917-8_8

Chapters 3, 4, and 5 show that tropical storms and related natural hazards have devastating impacts not only on the houses and the physical environment but also on the perception of people living in the affected areas. A first estimate of the costs of this damage is presented. The estimate goes beyond the direct payments by households and (local) authorities. Instead three types of capital which are relevant in a more comprehensive context for the households are assessed. The results show that first of all the physical capital is affected. Moreover, as natural hazards occur infrequently, significant differences exist over the years and among the most affected neighborhoods. The book provides a first estimate of the budget authorities need to set aside as compensation for natural hazard costs during the years to come.

This book shows that the main human ecological effects along the coast of Ky Anh (Vietnam) can be summarized as:

- People perceive extended periods of extreme temperatures, storms, and sea level rise as the most impacting environmental effects on their lives. In combination with drought, this results in less land used for rice paddies. In practice, farmers first turn rice fields in nonirrigated, sometimes pre-saline areas, for peanut cultures. Later on, as drought intensifies, the peanut areas are at risk of being abandoned, resulting in bare land. Impacting erosion of the sandy beaches and the dune ridge is widely present, in particular in the communities of Ky Anh. This increasingly affects the life in the local communities and the tourism potential of the area.
- Livelihoods and social damage are about people losing their house or forced to relocation. On the other hand, the effects on agro- and aquaculture left major imprints in the perception of the affected inhabitants. Post-hazard mental health problems, including the post-traumatic stress disorder (PTSD), have been described both in adults and in children (De Soir et al. 2014). In particular it is well known that following natural hazards and acute environmental incidents, the need for professional assistance on mental health increases. A study of PTSD effects and their cost in Ky Anh was beyond the scope of this research.
- Noticeable impacts are related to the cost of storms and related natural hazards. The perception study showed that the physical alterations caused by natural hazards affect the property and to a lesser extent, the income of people. The results show natural hazards cause annually 0.26 million $US damage of direct effects and the trend increases. However the indirect cost of the relocation policy and the brain drain caused by out-migration are currently not quantified but likely much higher. The assessment of three livelihood capitals shows that mainly the natural and the physical capitals of the communities are directly affected.

8.1.2 Risks for Nature and Humans in Uplands

Chapters 6 and 7 show that increasing pressures emerge on local livelihoods and slope agriculture as compared to the past. Climate change hazards are considered as the main pressure on the slope agriculture; especially the effects of heavy rains, flash floods, and floods are serious. Socioeconomic factors hardly put agriculture under pressure in the mountains. Land use policies and agricultural production aim at reducing the negative impacts of both the biophysical and the socioeconomic factors.

Particularly in the Van Chan mountainous district, it is most obvious that multi-hazards as the combination of heavy rains, flash floods, floods, and landslides are common and result in vast damage in local livelihood and agricultural production during the rainy season. In 2005, one of the biggest ever landslides on the hill next to the Son Thinh mountain resulted in an extensive degradation of the land. Only a limited area was upgraded afterward and is currently used for livestock and agriculture. The major part remained unused bare land until now.

8.2 Adaptation Policies and Measures

In spite of the inherent limitations associated with the preliminary character of this study, recommendations emerge:

8.2.1 Coastal and Beach Protection

Beaches support tourism along the Vietnamese coast. However coastal erosion is impacting beaches all over the major coasts of the Red River Delta, the central region, and the Mekong Delta of Vietnam. Protecting the beach resorts is a major challenge, since they illustrate prominently the balance between the economic and environmental aspects. Innovative, smart, and wise solutions protecting the beach should be applied (Nguyen et al. 2018).

An integrated approach combining natural, technical, and social knowledge, implemented through partnerships and coordination across disciplinary boundaries is crucial for risk management (Bernatchez and Fraser 2012). Ky Anh offers an example of Vietnam wide and international importance of the need for coastal protection and regional planning. Although Vietnam has a long-standing and effective history of coastal protection and flood management, the intensified storm and monsoon conditions offer new challenges. Evidence increases showing that a technical approach alone will likely prove insufficient handling the complex storm effects (WB 2010; Nguyen et al. 2018). Currently the district and the province fortify dikes and plant-protecting tree ridges and restore and extent mangroves (Fig. 8.1).

8.2.2 Green Cities

The human ecological effects of natural hazards and extreme (weather) events are complex and interrelated. An analytical approach addressing in a systematic way the different effects and aspects of the issue (e.g., by technological and engineering means) will therefore be insufficient. It should be complemented by more integral, integrated, and holistic actions.

Fig. 8.1 Both NGOs and Vietnamese Government increased the investment budget for the restoration of mangroves and their extension along the Ky Anh coast to protect the coastline. (Source: An Thinh Nguyen 2013)

Increasing attention during recent years is attracted by the "green cities" strategy. The concept emerged from the challenge of turning the weaknesses of postindustrial cities (pollution, urban degradation, consumptive resource use, loss of livability) into opportunities. They focus on a sound environment in which accessible green and respect for the natural resources is a main, but definitely not the only component.

Green cities have a strong climate change prevention component. Their advanced environmental management, planning, and policy address the choice of their sustainable energy sources, reducing the energy consumption, and prevent spillage and overconsumption of energy. They go for carbon neutrality and in the longer run for zero carbon dioxide emissions (Lucarelli and Roe 2012). They integrate in their planning cost-effective adaptation and counteraction strategies to the increasing frequency and intensity of storms and sea level rise.

However their targets are wider:

- Green cities go for "zero waste" (in which all waste is used as a resource), a clean surface and outlook, high quality of piped drinking and surface water, and the optimal use of the soil.

- Socially cities offer a healthy environment, respond to changes in their demography (as an aging population and the integration of migrants), and provide an equitable (income, opportunities) place to live.
- Cities deal with a fast-changing economy. They face main challenges in economic restructuring: growing income inequality and bipolar (high wages for experts versus low skills and wages jobs) labor markets to list just these examples.

The concept of the green city dovetails in its predecessor, the healthy city in which health was defined in the wider World Health Organization (WHO) context which goes beyond the absence of disease, but relates to mental and physical well-being.

To realize its ambitious targets, the green city as a rule uses "smart," often information technology (IT)-driven instruments. Smart city incentives deal with the efficient use of natural resources and energy, transport and mobility, buildings, and the quality of life, government, economy, and people. "Smart" in this context means innovative (often driven by advanced IT application but also using green infrastructure), skilled, inclusive, and sustainable (Neirotti et al. 2014; Nguyen et al. 2017).

The "green city" idea focuses on urban developments, while many climate change associated challenges of the human environment are also localized in remote coastal areas or in the mountains. Therefore the green city idea covers only part of the problems. Nevertheless the accent on urban ecosystems is legitimate for a series of reasons:

- Since 2008, over 50% of the population worldwide lives in cities (UN Habitat 2009). The urban metabolism consumes 65% of the physical resources (food, energy, water) which are attracted from outside their territory. Cities consume 75% of the world's energy and produce 80% of the greenhouse gas emissions. They have a pivotal role in counteracting climate change associated effects.
- Many important cities are located at the edges of the continents. Most of them in low-lying areas, which makes them most vulnerable for the effects of sea level rise, (tropical) storms, and the associated floods. This applies in particular to environmental problems requiring an adapted and tailored management but also has a dominating function in the social life and the economy of their hosting cities, today and in the future (Tran et al. 2014).
- Although scale matters for green cities, the basic ideas of sound, proactive, and smart environmental management can be applied locally. Eco-neighborhoods, eco-parks, eco-hotels, and eco-shopping centers show an increasing appeal and commercial success worldwide.

8.2.3 Sustainable Land Use, Integrated Land Use Planning, and Resilient Housing and Landscapes

In the lowland areas, the storms and related natural hazards affect land use change and land use planning in different ways. This necessitates a policy replying to the variety of treats for the local population. Drought requires according to the farmers more irrigation and crop selection changes. The income from forest products increases by developing agroforestry, diversifying the canopy, and reforestation, in particular on abandoned and bare land. Floods should be dealt with by fortifying and extending the dikes and by planting tree zones protecting the hinterland from the storm originating over the sea. Livestock can be secured by relocating the breeding farms to higher areas, which also provides advantages dealing with extreme high temperatures. Aquaculture might increase its profitability by restoring and planting mangroves in which a more diversified spectrum of fishes and seafood should be grown. Developing industry and tourism diversify the economy and make the area less dependent on agriculture alone.

Tran et al. (2015) suggested that Vietnam should invest in "storm-resilient housing," especially in Central Vietnam. Immediately after a destructive storm, authorities need to be prepared offering the victims at least a place to stay. These were community houses and tents along coast. In the longer term, households are offered new houses 2 to 3 kilometers land inward, built partially with government support. Sections of the new houses have already been offered to the storm victims. People are attached to the place where they and their family lived, sometimes for generations. The new neighborhoods are an important more sustainable asset.

More in particular the conclusions of the study on the Ky Anh coast (part 2) can be used for decisions on *climate change adaptation*, while they contribute to *sustainable development* and *sustainable adaptation*. Integrating local knowledge in adaptation responses is expected resulting in more resilience of and sustainability for the local community.

In the upland areas (part 3), this book supports the following management recommendation and policy lines for agricultural land use planning and land management in the Vietnamese northern mountains. Maintain indigenous knowledge to grow crops on the slopes, and adapt climate change hazards and land management policies in Vietnam. Mainstreaming traditional cultivation practices in land use polices was suggested in different countries to improve the effectiveness of land management (Folving and Christensen 2007; Rerkasem et al. 2009). Integrating traditional knowledge in all stages of land use decision-making and management is essential (Abule et al. 2005; Moges and Holden 2007; Reed et al. 2007). In practice, farmers respond to climate change hazards in different ways depending on the specific local conditions (Le and Tuan 2004; Quang et al. 2014). These differences are related to the biology and the physical aspects of their fields. Differences in agricultural methods exist among ethnic minorities and resulted in different land use models (Subedi et al.

2009; Misbahuzzaman 2016). Attitudes and knowledge of ethnic people are crucial factors of land use in defined situations (Hurni 2000; Zurayk et al. 2001; Gray and Morant 2003; Styger et al. 2007).

On the Van Chan mountain in particular and on other mountains in general, indigenous knowledge and skills are important for agriculture because they show how the indigenous farmers deal with the climate change hazard impacts on their fields. This knowledge was built up during the history of the local populations and is applied in combination with new cultivation technologies in the fields. Traditional methods are important instruments for farmers in this area: they allow farmers not only to increase the yields but also to protect soils. The opinion of the indigenous farmers allows providing core information about sustainable agricultural land use. Therefore, indigenous knowledge should be taken into account as it is more effective for agricultural land use planning in the Vietnamese mountains.

Overall this book contributes to ecosystem-based management, sustainability, and climate change resilient goals on local socioeconomic development and landscape planning. The study results in both lowland and upland of Vietnam advocate developing sustainable ecosystems (mangroves, coastal protected forests, climate change resilient crops, and mixed crop fields), upgraded new rural planning (NRP) (mainstreaming climate change hazard adaptation in the NRP process), and renewable energy strategies as the main local adaptations to climate change hazards.

8.2.4 Prospective Vision and Actions

In spite of the important budgets authorities spend dealing with the direct consequences of the climate change hazards, and the even more important investments they plan in the framework of a relocation policy, they have insufficient funds to support public programs protecting the coastal communities. As a result, coastal communities face considerable challenges on managing their responses to anticipate the damage from future natural hazards in the context of climate change (WB 2010). In the Vietnamese Central coast, the reply currently provided by the authorities is twofold:

8.2.4.1 The Military Road

The military road strengthens the physical barriers protecting the hinterland from the effects of natural hazards. Along the Vietnamese Central Coast, the military roads act as a dam just behind and parallel to the beaches. This protection structure should be fortified. In Vietnam, as in many places elsewhere, it is common practice finding roads built on dams, protecting the hinterland. The military road was designed preventing both flooding and salt intrusion. It could not stand the hard conditions of contemporary storms. Coastal areas invest in tree planting projects in the dunes.

8.2.4.2 The Layer Model of Storm Management

Based on the results of the household questionnaire about the post-hazard support in the Ky Anh coast (Chap. 3), it is indicated that the local government provides the hazard information timely allowing local people to prepare for and respond to the hazard. In this area one might also apply the Dutch model of climate change hazard management. This layer model analyzes, designs, and organizes participation in planning land use, water management, and civil engineering works as well as in the communication on these matters.

Dijkman (2007) introduced a three-layer model:

- Layer 1: The *underground layer* (soil, water, flora, and fauna) is crucial for sustainability.
- Layer 2: The *network layer* consists of all forms of infrastructure (waterways, roads, railways, levees, sluices, locks, etc.).
- Layer 3: The *occupation layer* consists of spatial patterns of buildings for human use.

Measures of layers 2 and 3 are indicated to deal with the damage by natural hazards. Changes of the network layer spread over 25 to 100 years, while changes of the occupation layer need periods of 10 to 25 years.

8.3 National Challenges and International Cooperation

8.3.1 National Challenges and Local Effects

If the sea level increases by 1 meter, 3% of the land along the Vietnamese coast will be inundated, and about 10–12% of Vietnam's population is directly impacted. Consequently, the country will lose an estimated 10% of its GDP. The effects are associated with poverty, the realization of millennium goals, and the sustainable development of Vietnam. These most impacting effects explain why many national documents deal with climate change impacts recently.

In Vietnam, the first important policy related with climate change is resolution number 24/NQ/TW of the government on climate change improvement on natural resources management in 2013. It was the direction of almost all climate change activities. The overall objectives of the resolution are:

- By 2020, to basically adapt to climate change, prevent natural hazards, reduce greenhouse gas emissions, and make fundamental changes in the exploitation and use of natural resources in a rational, effective, and sustainable manner, limiting the increase of environmental pollution and biodiversity loss in order to ensure the quality of the living environment, maintaining the ecological balance, toward a green economy, which is friendly for the environment.

- By 2050, actively respond to climate change, using rational, economical, effective, and sustainable resources; ensure the quality of the habitats and the ecological balance; and strive to achieve environmental targets equivalent to the current levels prevailing in the industrialized countries in the region.

The Law on Economical Energy Efficiency Use was launched in 2010 to guide the energy efficiency use in different economy sectors, to reduce the energy use, to save the environment, and to reduce the impact on climate change.

In 2011, Vietnam approved the National Strategy on Climate Change until 2020, which is based on ten strategic goals.

By 2012, the National Commission on Climate Change was established. The commission deals with:

- The integrated policy, strategies, and programs, at medium and long term for climate change adaptation in Vietnam.
- The expected climate change effects in sectors and addressing the major problems.
- The issues related to climate change in the national policies, programs, and projects.
- Participation in important international cooperation projects and international agreements on climate change.
- Preparation of the legal documents coping with climate change.

The Law on Natural Hazard Prevention and Control was enforced in 2013. The law addressed the activities on natural hazard prevention, protection, and responses to reduce the negative impacts of the natural hazards.

In 2014, the Law on Environment Protection has been revised and expanded with a chapter dealing with climate change. The general regulation on climate change has been established; particularly integration of a climate change strategy, in planning, and socioeconomic development plans was made compulsory. The regulations on greenhouse gas emissions and the management of ozone layer depleting substances were provided with proper management tools. Energy efficiency use is a core part of the chapter. It indicates the attention for climate change issues by the government.

In 2015, the Law on Hydrology and Meteorology was enforced. One chapter focuses on climate change monitoring, data collection, national meteorology assessment, and integration of climate change scenarios and climate change monitoring in the water resources planning.

A series of national strategies such as the National Strategy on Hazard Prevention and Control (2007), National Climate Change Strategy (2011), Vietnam's Green Growth Strategy (2012), and Renewable Development Energy Strategy (2015) have been developed giving the more detailed directions for management and development of the country on climate change. Moreover, priority adaptation projects and programs, co-benefit projects and programs, and priorities mitigation projects and programs were frequently approved by the government. The National Target Program to Respond to Climate Change (NTP-CC) has been developed by that time. In order to mobilize fund from development partners for the NTP-CC, the

Support Program to Respond to Climate Change (SP-RCC) has been established. Under SP-RCC, policies have been developed by the competent ministries from 2009 to 2017 giving more clear guidelines for management and development.

Human ecological effects of natural hazards stress the need for laws, strategies, programs, and projects mentioned above at a local scale. Applying policies regionally and locally is currently the most important response to climate change adaptation and natural hazard mitigation in Vietnam.

8.3.2 International Cooperation on Climate Change, Local Adaptation, and Hazard Prevention

The Vietnamese Government joined and approved the United Nations Framework Convention on Climate Change (UNFCCC) and the Kyoto Protocol. Vietnam as a party in the convention systematically follows up the implementation conferences of the parties as the Stockholm and Paris ones. Its policy focuses on the natural hazard prevention and mitigation. Locally, international cooperation projects are often based on enhanced engineering. International cooperation does not appeal to local authorities. Local governments should target financial and technological support from the international community on natural hazard mitigation, climate change adaptation, and resilience. This might significantly alleviate the existing problems local authorities currently face.

Examples are implemented in the Ha Tinh province. In 2012, the province administration formulated the management and implementation terms of reference of the project *essential infrastructure development for coastal mudflats to deal with climate change impacts in the Ha Tinh province"* with the financial support of the OPEC Fund for International Development (OFID). The project funded by the Belgian Development Agency *integrated water management and urban development in relation to climate change in the Ha Tinh province* (2013–2020) linked goals of water resources management and urban development with climate change issues. The governments of Ky Anh and its coastal communes need to cooperate with overseas countries and international organizations in preventing and controlling environmental pollution, preserving nature and biodiversity, and coping with climate change.

8.4 Limitations of the Book

8.4.1 Uncertainties of Perception Studies Using a Likert Scale

Human ecological research on the effects associated with climate change is inherently characterized by scientific uncertainties. A specific uncertainty related with the results described in this book is associated with the perception analysis (Chaps. 3, 4, 5, and 7). Overall the climate change hazards are hardly mentioned by the interviewees. In particular for the items considered of limited importance (livestock, income, etc.), this might be the result of bias. This book ranks effects of the hazards according to people's perception based on mean values in a Likert scale. The Likert scale is widely used for this type of research and is considered an effective tool in many disciplines. However, the Likert scale is known having main limitations: more complex relationships are hard to assess in this way, while the respondents are unaware of this limitation (Hartley 2014).

Chapter 7 deals with mountains in Northern Vietnam because these areas are most affected by both natural and human impacts. Consequently all selected respondents are "experts by experience," in particular because the focus of the study is on local impacts and responses. Therefore, perceptions of farmers reflect the present pressures, state, and responses to sustainability of their local livelihood and agriculture on the slopes.

8.4.2 Reliability of the Delphi Survey

Overall Delphi techniques are recognized as complementary analysis tools in the international literature (Linstone and Turoff 2002; Mukherjee et al. 2015; Le et al. 2015; Ballantyne et al. 2016). Nevertheless one might use the results for making decisions in practice. The consensus among the stakeholders is an important element for planning (Linstone and Turoff 2002; Hsu and Sandford 2007; Njoroge 2014). However the reliability of the Delphi (Chaps. 5 and 7) results largely depends on the selection of the panel members, the open questions of the preliminary stage, using communication techniques adapted to the specific needs, and addressing problems the panel members are familiar with during different rounds (Hsu and Sandford 2007). Particularly a case study of the Ky Anh coast (Chap. 5) introduces strategies to handle these problems:

- First, a panel of local people mixed with local decision-makers was established because a consensus among the stakeholders is an important element for local decision-making (Lund et al. 2014). Eight representatives of the local authorities were selected in seven communes (two persons in Ky Xuan). This reflects the current situation in Vietnam, in which two administration officials are competent for natural resources and the environment in the communes. Statistics and documents provided by the Ky Anh district government allowed selecting 28 local citizens of 20 villages most seriously affected by natural hazards. They originate from small communities (about 15–30 households per village) and live along the coastline, and their income depends to a large extent on small-scale farming and fishing. A total of 36 panelists is the required sample size for a Delphi study (between 15 and 60 participants) (Hasson et al. 2000).
- Second, the preparatory questionnaire: the questions of the first round are based on the DPSIR framework, which allows stakeholders understanding easily climate change associated hazards in terms of drivers, pressures, states, impacts, and responses.

- Third, different communication techniques are used in Delphi approaches, mail, email, telephone, or face-to-face interview, to list just these instruments (Gupta and Clarke 1996; Zolingen and Klaassen 2003; Le et al. 2015). This study uses face-to-face interviews in both Delphi rounds because the panelists prefer direct contacts. This option limited the dropout of the participants between the two rounds and contributed to the high reliability and relevance of the results.

This study resulted in a *strong agreement* and *high confident* consensus among the local communities and local authorities on climate change impacts and adaptation after two Delphi rounds. Rowe and Wright (1999) showed that ideally a Delphi process should be concluded after three to five rounds. For example, the HighARCS project in upland Vietnam (Lund et al. 2014) used a closed questionnaire in round 3 following an insufficient consensus between the panel members after the second round. This study needed only two rounds because of these following reasons:

- In contrast to aspects of socioeconomic development, environmental pollution, climate change, and climate change hazards are low policy priorities in Vietnam today.
- This study deals with coastal areas of Central Vietnam because this area is most affected by storms, and consequently all panelists are "experts by experience" in particular because the focus of the study is on local impacts.
- Both local inhabitants and authorities easily understood the questions and also were most willing to reply.

8.4.3 Uncertainties on Damage Valuation

The case study of the Ky Anh coast, in Chap. 4, shows storms have devastating impacts not only on the houses and the physical environment but also on the perception of people living in the affected areas. A first estimate of the costs of this damage is presented. The estimate goes beyond the direct payments by the households and the (local) authorities. Instead three types of capital which are relevant in a more comprehensive context for the households are assessed. The results show that first of all the physical capital is affected. Moreover, as storms occur infrequently, significant differences exist over the years and among the most affected neighborhoods. Chapter 4 provides information on the budget authorities need to set aside compensating for storm costs during the years to come.

Nevertheless monetization also shows a set of uncertainties:

- The estimations of the costs caused by the storms are based on data provided by local authorities and the households. These data mainly cover direct costs and tend externalizing other aspects of the different capitals under study.
- The most significant yearly variations in storm frequency and intensity show that yearly budgets alone (as e.g., used by authorities) are not the most appropriate instrument dealing with compensations for storm impacts. Medium- and long-term budget instruments (funds or bonds) should be envisaged. Nevertheless, combining the existing data, a medium-term increase in the storm abatement costs is evident.

- The forecasting period runs until 2019 (6 years). Even when this period is limited, predictions have a limited confidence. On the other hand, it is the best estimate one might currently provide for a most relevant policy question. As this type of data will accumulate in the future, better, more confident, and reliable forecasts will become available.
- Overall, the monetary aspects in this study should be considered as preliminary. The research is based on measured data. The more case studies as the ones presented here become available, the more accurate and reliable answers one might provide to the legitimate policy question: *How much do climate change (mediated) effects cost?*

The NPV is considered a significant indicator of hazard damage and particularly allows dealing with decisions on storm risk prevention (Wilks 2013). Globally, a loss of 400 billion $US caused by natural hazards during the 1990s was calculated; this figure could be reduced to 280 billion $US with a 40 billion $US investment in natural hazard preparedness, prevention, and mitigation (Dilley and Heyman 1995). Vietnam faces storms with the highest cost bill in Southeast Asia (Shreve and Kelman 2014). For the Ky Anh coast, this study calculates the estimated monetary cost of risk reduction as 1.19 up to 1.32 million $US for the next 6 years (2014–2019). To the best of our knowledge, till now no study deals with the valuation of the cost of storms at the local scale (district and commune) in Central Vietnam.

8.4.4 Added Value of the Book

This book uses a human ecological approach to study effects of natural hazards on the nature and humans at national, regional, and local scales. This approach is not unique. It builds on previous research, e.g., by the group of Adger and his co-workers: Social vulnerability to climate change and extreme weather conditions (Adger 1999), adaptation to climate change (Adger et al. 2003), climate change-related human security and violent conflicts (Barnett and Adger 2007), changing social contracts in climate change adaptation (Adger et al. 2013a, 2013b), cultural dimensions of climate change impacts and adaptation (Adger et al. 2013a, 2013b) are all subjects addressed by this team.

Nevertheless from a human ecological viewpoint, the book provides specific added value on a series of aspects.

- *Climate Change Associated Facts*

There is very little, if any, room for doubt left. An unmistakable rise in climate change associated effects exists worldwide. This book illustrates this statement with a multitude of research facts for Vietnam. Storms, floods, sea level rise, and coastal erosion are most convincingly illustrated. The perception of the directly involved stakeholders is inventoried, and a first estimation of different types of direct costs is provided. The over 3000-kilometer-long coast of Vietnam and the concentration of its economy in the deltas and along the coastal area make the country particularly vulnerable for pressures on both nature and the anthropogenic environment. Along only few beaches and ports worldwide, the beach erosion and storm damage responses are as pronounced and tangible as in Vietnam.

- *Research Strategy*

This book provides a convincing example on how biophysical effects of climate changes (storms, landslides, floods) are linked with the effects on individuals, families, and communities. Their perception and (traditional) knowledge provides an important basis for mitigation, adaptation, and policy intervention. The book shows that unraveling this complexity necessitates the targeted combination of a series of research approaches and methods regularly applied in both natural and human sciences. The way these methods are combined and the interrelated results they provide can be used as a template for future human ecological research.

- *The Coevolution Between Local Communities and Their Changing Environment*

The relationships between local communities and their environment which is subject to climate change hazards are described in case studies. This book is about the impacts of climate change hazards on local communities and focuses on the importance of changing human behavior to cope with these hazards both along the coast and in the highlands of Vietnam. Its transdisciplinary approach contributes to a more in-depth understanding of the human factors in the changing environments.

- *Evidence About the Connection of Sustainable Development in the Upland and the Lowland*

In general much less information on the impact of agriculture in the mountains is available as compared to the coast and its direct hinterland. The protected forests and crop systems on the slopes of upland Vietnam are particularly important both locally and regionally. From a landscape ecological viewpoint, mountains (central, central highland, and northern mountains) are the "sources" which directly affect the economy and the environment in the "sink" of the lowlands (Red River Delta, central coast). Therefore, in a global context of climate change, sustainable development in the mountains is connected with the development of the lowlands in the deltas and near the sea.

- *The Uncertainty.*

In spite of the wide and interdisciplinary approach of this book, many aspects of climate change associated effects remain out of its scope. Droughts and its associated effects as fires is an increasing problem in particular in the southern provinces; impacts on soil, impacts on mangrove ecosystems and on water, and food security are most real in Vietnam but remain out of the scope of this book.

- *The Timely Character of the Book*

At the moment the text of this book manuscript is concluded, early November 2017, the tropical storm Damrey hits Central Vietnam. The storm anchored the mainland close to one of the study areas (Ky Anh) of this book. The accompanying winds, intense rains, and floods caused deaths, injuries, and damaged houses and infrastructure, as many of the storms which hit Vietnam during recent years (for an overview, see Chap. 4), and strengthen the trends initiated before. In this context, this book is a repeated wake-up call for those denying the effects of climate changes and the most necessary context to mitigate, stabilize, and alleviate them. On the previous pages, ample data and inspiration are provided on how such a policy should like for individuals, households, communities, and districts.

8.5 Conclusion

This book dissects, relates, and integrates the effects of natural hazards (in the context of climate change) as storms and landslides on the biophysical and the socio-economic environment of the inhabitants of a heavily affected coast in Central Vietnam and the mountains in the north of Vietnam. This concluding chapter aims at echoing the core issues at stake and highlights their policy relevance.

The text takes off with a summary of the risks which are at the same time comparable and different in both geographic regions. Based on this information, the text slides into the question "How these deleterious effects can be mitigated?" Nature-bound beach protection, green cities, integrated land use planning, increased resilience, and an inspired visionary action on the future emerge as main options.

Climate change policy needs to be implemented from the very local to the international level. A review of the national Vietnamese policy documents on climate change shows that the country is a loyal partner in the international policy debate driven by the implementation of the UNFCCC. Consequently the country put a comprehensive and multifaced legal framework in place to deal with climate changes, including the natural hazards which are at the core of this book.

Human ecological research in particular on a subject as climate changes is not only characterized by hard facts but also by scientific uncertainty. Three main areas of uncertainty related to the research in this book are identified:

- The structure of the perception studies in which the selection of the participants is key to define their scope.
- The reliability of the Delphi surveys characterized by the quality of the questionnaires and the used communication techniques.
- The monetary damage valuation which (mainly) addresses the direct costs of the impacts but externalizes important social costs.

A concluding section entails reflections on the specific added value of the human ecological approach, which is an important aspect of the specificity of this book. Attention is given to the research strategy which hopefully will provide inspiration and can be used as a template for similar research in other countries and regions. Also the coevolution of local communities and their environment, both along the coast and in the highlands, emerges in this discussion.

References

Aalst, M. K., Cannon, T., & Burton, I. (2008). Community level adaptation to climate change: The potential role of participatory community risk assessment. *Global Environmental Change, 18*(1), 165–179.

Abule, E., Snyman, H. A., & Smit, G. N. (2005). Comparisons of pastoralists perceptions about rangeland resource utilization in the Middle Awash Valley of Ethiopia. *Environmental Management, 75*, 21–35.

Acharya, G. P., Tripathi, B. P., Gardner, R. M., Mawdesley, K. J., & Mcdonald, M. A. (2008). Sustainability of sloping land cultivation systems in the mid-hills of Nepal. *Land Degradation and Development, 19*(5), 530–541.

Adger, W. N. (1999). Social vulnerability to climate change and extremes in coastal Vietnam. *World Development, 27*(2), 249–269.

Adger, W. N., Huq, S., Brown, K., Conway, D., & Hulme, M. (2003). Adaptation to climate change in the developing world. *Progress in Development Studies, 3*(3), 179–195.

Adger, W. N., Barnett, J., Brown, K., Marshall, N., & O'Brien, K. (2013a). Cultural dimensions of climate change impacts and adaptation. *Nature Climate Change, 3*, 112–117.

Adger, W. N., Quinn, T., Lorenzoni, I., Murphy, C., & Sweeney, J. (2013b). Changing social contracts in climate-change adaptation. *Nature Climate Change, 3*, 330–333.

Adriaanse, A. (1993). *Environmental policy performance indicators: A study on the development of indicators for environmental policy in the Netherlands*. The Hague: SDU Publishing, 175 pages.

Aguirre, B. E., Saenz, R., Edmiston, J., Yang, N., Agramonte, E., & Stuart, D. L. (1993). The human ecology of tornadoes. *Demography, 30*(4), 623–633.

Alesheikh, A., & Nouri, N. (2007). Coastline change detection using remote sensing. *Environmental Science and Technology, 4*, 61–66.

Allen, I. E., & Seaman, C. A. (2007). Likert scales and data analyses. *Quality Progress, 40*(7), 64–65.

Armenteras, D., Espelta, J. M., Rodríguez, N., & Retana, J. (2017). Deforestation dynamics and drivers in different forest types in Latin America: Three decades of studies (1980–2010). *Global Environmental Change, 46*, 139–147.

Avinash, K., Narayana, A. C., & Jayappaa, K. S. (2010). Shoreline changes and morphology of spits along southern Karnataka, west coast of India: A remote sensing and statistics-based approach. *Geomorphology, 120*, 133–152.

Ballantyne, R., Hughes, K., & Bond, N. (2016). Using a Delphi approach to identify managers' preferences for visitor interpretation at Canterbury Cathedral World Heritage Site. *Tourism Management, 54*, 72–80.

A. T. Nguyen, L. Hens, *Human Ecology of Climate Change Hazards in Vietnam*, Springer Climate, https://doi.org/10.1007/978-3-319-94917-8

Bari, S. H., Rahman, M. T., Hussain, M. M., & Ray, S. (2015). Forecasting monthly precipitation in Sylhet city using ARIMA Model. *Civil and Environmental Research, 7*(1), 69–77.

Barnett, J., & Adger, W. N. (2007). Climate change, human security and violent conflict. *Political Geography, 26*(6), 639–655.

Benitez-Capistros, F., Hugé, J., & Koedam, N. (2014). Environmental impacts on the Galapagos Islands: Identification of interactions, perceptions and steps ahead. *Ecological Indicators, 38*, 113–123.

Bernatchez, P., & Fraser, P. (2012). Evolution of coastal defence structures and consequences for beach width trends, Québec, Canada. *Coastal Research, 28*(6), 1550–1566.

Bettini, G. (2013). Climate Barbarians at the Gate? A critique of apocalyptic narratives on "climate refugees". *Geoforum, 45*, 63–72.

Biloslavo, R., & Dolinšek, S. (2010). Scenario planning for climate strategies development by integrating group Delphi, AHP and dynamic fuzzy cognitive maps. *Foresight, 12*(2), 38–48.

Biloslavo, R., & Grebenc, A. (2012). Integrating Group Delphi, Analytic Hierarchy Process and dynamic fuzzy cognitive maps for a climate warning scenario. *Kybernetes, 41*(3/4), 414–428.

BKG (Bac Kan Government). (2009). *Report on natural hazard prevention 2008*. Bac Kan, 20 pages (in Vietnameses).

Boak, E. H., & Turner, I. L. (2005). Shoreline definition and detection: A review. *Coastal Research, 21*(4), 688–703.

Box, G. E. P., & Jenkins, G. M. (1976). *Time series analysis: Forecasting and control* (2nd ed.). San Francisco: Holden-Day, 575 pages.

Box, G. E. P., Jenkins, G. M., & Reinsel, G. C. (1994). *Time series analysis, forecasting and control* (pp. 197–199). Engle Wood Cliffs, NJ: Prentice Hall.

Buhlmann, E., Wolfgramm, B., Maselli, D., Hurni, H., Sanginov, S. R., & Liniger, H. P. (2010). Geographic Information System-based decision support for soil conservation planning in Tajikistan. *Soil and Water Conservation, 65*(3), 151–159.

Bunting, S. W. (2008). Horizontally integrated aquaculture development: Exploring consensus on constraints and opportunities with a stakeholder Delphi. *Aquaculture International, 16*(2), 153–169.

Butze, K. W. (1983). Human response to environmental change in the perspective of future, global climate. *Quaternary Research, 19*(3), 279–292.

Carter, R. W. G. (1987). Man's response to sea-level change. In R. J. N. Devoy (Ed.), *Sea Surface Studies* (pp. 464–498). Springer: Netherlands.

CCFSC. (2006). *Handbook on damage assessment of natural hazards*. Hanoi, Vietnam: MARD, 43 pages (in Vietnamese).

CCFSC (Central Committee for Flood and Storm Control). (2001). *Second national strategy and action plan for disaster mitigation and management in Viet Nam – 2001 to 2020. Ministry of Agriculture and Rural Development*. Vietnam: Hanoi, (in Vietnamese).

CCFSC (Central Committee for Flood and Storm Control). (2010). *Summary of damages caused by floods, flashfloods and landslides and result of control and prevention methods in the previous years (materials for online conference)*. Vietnam: Hanoi, (in Vietnamese).

CCFSC (Central Committee for Flood and Storm Control). (2014). *The report of damage caused by floods, flash floods, landslides and preventive measures in recent years, Documentation for the online conference in August 20, 2014 of the CCFSC*. Vietnam: Hanoi, (in Vietnamese).

Chambers, R., & Conway, G. R. (1992). *Sustainable rural livelihoods: Practical concepts for the 21st century. Discussion paper 296*. Brighton: Institute of Development StudiesBrighton, 29 pages.

Chen, L. C., & Rau, J. Y. (1998). Detection of shoreline changes for tideland areas using multitemporal images. *Remote Sensing, 1*(17), 3383–3397.

Chen, J. Y., Li, D. J., Yao, Y. M., Li, J., & Jin, W. H. (1999). Marine hazards in coastal areas and human action against them – A case study of the Changjiang River estuaries and Qiantang River. *Chinese Journal of Oceanology and Limnology, 17*(2), 143–153.

Choi, D., Jun, H. D., & Kim, S. (2012). *Vulnerability assessment of natural disasters for small and mid-sized streams due to climate change and stream improvement*. EGU General Assembly 2012, held 22–27 April, 2012 in Vienna, Austria, p. 4060.

Colagiuri, R., Boylan, S., & Morrice, E. (2015). Research priorities for NCD prevention and climate change: An international Delphi survey. *Environmental Research and Public Health, 12*(10), 12941–12957.

Cruz, R. V., Harasawa, H., Lal, M., Wu, S., Anokhin, Y., Punsalmaa, B., et al. (2007). Asia: Climate change 2007: Impacts, adaptation and vulnerability. In M. L. Parry et al. (Eds.), *Contribution of working group II to the fourth assessment report of the intergovernmental panel on climate change* (pp. 469–506). Cambridge: Cambridge University Press.

Cui, F. (2011). *ARIMA models for bank failures: Prediction and comparison*. Las Vegas: University of Nevada, 57 pages.

Dasgupta, S., Lapla, B., Murray, S., & Wheeler, D. (2009). *Sea-level rise and storm surges: A comparative analysis of impacts in developing countries, World Bank Policy Research Working Paper (42), WPS4901*. Washington, DC: World Bank, 41 pages.

Delisle, S., & Turner, S. (2016). 'The weather is like the game we play': Coping and adaptation strategies for extreme weather events among ethnic minority groups in upland northern Vietnam. *Asia Pacific Viewpoint, 57*(3), 351–364.

De Soir, E., Zech, E., Alisic, E., Versporten, A., Van Oyen, H., Kleber, R., van der Hart, O., & Mylle, J. (2014). Children following the Ghislenghien gas explosion: PTSD predictors and risk factors. *Journal of Child Adolescent Trauma, 10*, 189–200.

DFID. (1999). *Sustainable livelihoods guidance sheets: Framework*, Section 2. Subsection 2.1, 25 pages.

Dijkman, J. (2007). *A Dutch perspective on coastal Louisiana flood risk reduction and landscape stabilization. Final Report. Netherlands Water Partnership* (p. 279). England: Netherlands. London.

Dilley, M., & Heyman, B. N. (1995). ENSO and disaster: Droughts, floods and El Nino southern oscillation warm events. *Disasters, 19*(3), 181–193.

Ding, Y. (2012). Sea-level rise and hazardous storms: Impacts on coasts and estuaries. In W. Y. Chen, J. Seiner, T. Suzuki, & M. Lackner (Eds.), *Handbook of climate change mitigation* (pp. 471–503). Springer.

Dohong, A., Aziz, A. A., & Dargusch, P. (2017). A review of the drivers of tropical peatland degradation in South-East Asia. *Land Use Policy, 69*, 349–360.

Doria, M. F., Boyd, E., Tompkins, E. L., & Adger, W. N. (2009). Using expert elicitation to define successful adaptation to climate change. *Environmental Science and Policy, 12*(7), 810–819.

Duncan, O. D., & Stenbeck, M. (1987). Are Likert scales unidimensional? *Social Science Research, 16*(3), 245–259.

EEA. (1995). *Europe's environment: The Dobris assessment*. Copenhagen: European Environmental Agency, 8 pages.

Eriksen, S., Aldunce, P., Bahinipati, C. S., Martins, R., Molefe, J. I., Nhemachena, C., O'brien, K., Olorunfemi, F., Park, J., Sygna, L., & Ulsrud, K. (2011). When not every response to climate change is a good one: Identifying principles for sustainable adaptation. *Climate and Development, 3*(1), 7–20.

FAO. (2012). World Agriculture: towards 2015/2030. http://www.fao.org/docrep/004/y3557e/y3557e08.htm

Farbotko, C., & Lazrus, H. (2012). The first climate refugees? Contesting global narratives of climate change in Tuvalu. *Global Environmental Change, 22*(2), 382–390.

Folving, R., & Christensen, H. (2007). Farming system changes in the Vietnamese uplands - using fallow length and farmers' adoption of Sloping Agricultural Land Technologies as indicators of environmental sustainability. *Geografisk Tidskrift, 107*(1), 43–58.

Gari, S. R., Newton, A., & Icely, J. D. (2015). A review of the application and evolution of the DPSIR framework with an emphasis on coastal social-ecological systems. *Ocean and Coastal Management, 103*, 63–77.

Graaff, J., Duarte, F., Fleskens, L., & Figueiredo, T. D. (2008). The future of olive groves on sloping land and ex-ante assessment of cross compliance for erosion control. *Land Use Policy, 27*(1), 33–41.

Granderson, A. A. (2014). Making sense of climate change risks and responses at the community level: A cultural-political lens. *Climate Risk Management, 3*, 55–64.

Gray, L. C., & Morant, P. (2003). Reconciling indigenous knowledge with scientific assessment of soil fertility changes in southwestern Burkina Faso. *Geoderma, 111*, 425–437.

GSO (Vietnamese Government Statistical Office). (2014). *Statistical yearbook of Vietnam 2013*. Vietnam: Statistics Publishing, 935 pages.

GSO (Vietnamese Government Statistical Office). (2016). *Statistical yearbook of Vietnam 2015*. Vietnam: Statistics Publishing, 947 pages.

Gupta, U. G., & Clarke, R. E. (1996). Theory and applications of the Delphi technique: A bibliography (1975–1994). *Technological Forecasting and Social Change, 53*, 185–211.

Hartley, J. (2014). Some thoughts on Likert-type scales. *International Journal of Clinical and Health Psychology, 14*(1), 83–86.

Hasson, F., Keeney, S., & McKenna, H. (2000). Research guidelines for the Delphi survey technique. *Advanced Nursing, 32*(4), 1008–1015.

HGG (Ha Giang Government). (2011). *Report on natural hazard prevention 2010*. Ha Giang, 9 pages (in Vietnameses).

Hsu, C. C., & Sandford, B. A. (2007). The Delphi technique: Making sense of consensus. *Practical Assessment Research and Evaluation, 12*(10), 1–8.

HTG (Ha Tinh Government). (2013). *Statistic Yearbook 2012*. Thong-ke Publisher, Vietnam, 452 pages.

Hugo, G. (2011). Future demographic change and its interactions with migration and climate change. *Global Environmental Change, 21*, S21–S33.

Hurni, H. (2000). Assessing sustainable land management (SLM). *Agriculture, Ecosystems & Environment, 81*(2), 83–92.

Imaduddina, A. H., & Subagyo, W. W. H. (2014). Sea level rise flood zones: Mitigating floods in Surabaya coastal area. *Procedia - Social and Behavioral Sciences, 135*, 123–129.

IPCC. (2001). Climate change 2001: Impacts, adaptation and vulnerability. In J. J. McCarthy, O. F. Canziani, N. A. Leary, D. J. Dokken, & K. S. White (Eds.), *Contribution of working group II to the third assessment report of the intergovernmental panel on climate change*. Cambridge, UK\New York, USA: Cambridge University Press, 1032 pages.

IPCC. (2007). Climate change 2007: Impacts, adaptation, and vulnerability. In M. L. Parry, O. F. Canziani, J. P. Palutikof, P. J. van der Linden, & C. E. Hanson (Eds.), *Contribution of working group II to the fourth assessment report of the intergovernmental panel on climate change*. Cambridge, UK: Cambridge University Press, 976 pages.

IPCC. (2014). *Climate change 2014 Synthesis Report*. Contribution of Working Groups I, II and III to the Fifth Assessment Report of the Intergovernmental Panel on Climate Change [Core Writing Team, R.K. Pachauri and L.A. Meyer (eds.)]. IPCC, Geneva, Switzerland, 151 pages.

IPONRE. (2009). *Ha Tinh assessment report on climate change*. Hanoi, Vietnam: Van hoa - Thong tin Publishing, 57 pages.

Jun, K. S., Chung, E. S., Kim, Y. G., & Kim, Y. (2013). A fuzzy multi-criteria approach to flood risk vulnerability in South Korea by considering climate change impacts. *Expert Systems with Applications, 40*(4), 1003–1013.

Jung, Y., Kim, D., Choi, M., Kim, S., & Park, M. (2014). Vulnerability resilience in the major watersheds of the Korean Peninsula. *Terrestrial, Atmospheric and Oceanic Sciences, 25*(6), 857–868.

KAG (Ky Anh Government). (2009). *Report on migration in Ky Anh 2008*. Ha Tinh. 12 pages (in Vietnamese).

KAG (Ky Anh Government). (2012). *Land use planning in Ky Anh district until 2020*. Ha Tinh, 128 pages (in Vietnamese).

KAG (Ky Anh Government). (2014). *Report on migration in Ky Anh 2013*. Ha Tinh, 05 pages (in Vietnamese).

Kaushik, I., & Singh, S. M. (2008). Seasonal ARIMA model for forecasting of monthly rainfall and temperature. *Environmental Research and Development, 3*(2), 506–514.

Kelly, P., & Huo, X. (2013). Do farmers or governments make better land conservation choices? Evidence from China's Sloping Land Conversion Program. *Forest Economics, 19*, 32–60.

Khuc, Q. V., Tran, B. Q., Meyfroidt, P., & Paschke, M. W. (2018). Drivers of deforestation and forest degradation in Vietnam: An exploratory analysis at the national level. *Forest Policy and Economics, 90*, 128–141.

Kim, Y., & Chung, E. S. (2013). Fuzzy VIKOR approach for assessing the vulnerability of the water supply to climate change and variability in South Korea. *Applied Mathematical Modelling, 37*(22), 9419–9430.

Kirezieva, K., Jacxsens, L., van Boekel, M. A. J. S., & Luning, P. A. (2015). Towards strategies to adapt to pressures on safety of fresh produce due to climate change. *Food Research International, 68*, 94–107.

Lane, D., Mercer, C. C., Forbes, D. L., & Watson, P. (2013). The gathering storm: Managing adaptation to environmental change in coastal communities and small islands. *Sustainability Science, 8*(3), 469–489.

Lawrence, R. J. (2003). Human ecology and its applications. *Landscape and Urban Planning, 65*(1–2), 31–40.

LCG (Lao Cai Government). (2009). *Report on natural hazard prevention 2008*. Lao Cai, 12 pages (in Vietnamese).

Le, D. Q., & Tuan, H. D. (2004). Improving indigenous technologies for sustainable land use in northern mountainous areas of Vietnam. *Mountain Science, 1*(3), 270–275.

Le, T., Nguyen, Q. T., Nguyen, M. T., Dang, D. L., Nguyen, V. P., Phi, C. V., Bui, X. D., Nguyen, D. V., Vu, V. C., & Thanh, N. L. (2012). *Vietnam: Land and people*. Hanoi, Vietnam: Vietnam Education Publishing House, 544 pages.

Le, T. H., Nguyen, A. T., Tran, A. T., Dao, D. C., Luu, T. A., Hoang, L. T. T., Nguyen, M. H., Tran, Q. B., Le, V. H., Uong, D. K., Bui, T. M., Tong, P. T., Hoang, H., & Truong, Q. H. (2015). Impacts of climate change on agro-ecological landscapes in the coastal area of the Thai Binh province (Vietnam) using the Delphi technique. *Climate Change Strategies and Management, 7*(2), 222–239.

Lemieux, C. J., & Scott, D. J. (2011). Changing climate, challenging choices: Identifying and evaluating climate change adaptation options for protected areas management in Ontario, Canada. *Environmental Management, 48*(4), 675–690.

Lemieux, C. J., Gray, P. A., Douglas, A., Nielsen, G., & Pearson, D. (2014). From science to policy: The making of a watershed-scale climate change adaptation strategy. *Environmental Science & Policy, 42*, 123–137.

Lewison, R. L., Rudd, M. A., Al-Hayek, W., Baldwin, C., Beger, M., Lieske, S. N., Jones, C., Satumanatpan, S., Junchompoo, C., & Hines, E. (2016). How the DPSIR framework can be used for structuring problems and facilitating empirical research in coastal systems. *Environmental Science and Policy, 56*, 110–119.

Li, K., & Li, G. S. (2013). Risk assessment on storm surges in the coastal area of Guangdong province. *Natural Hazards, 68*(2), 1129–1139.

Likert, R. (1932). A technique for the measurement of attitudes. *Archives of Psychology, 140*, 1–55.

Linstone, H. A., & Turoff, M. (2002). *The Delphi method: Techniques and applications*. Addison-Wesley Publisher, 620 pages.

Lucarelli, M., & Roe, P. G. (Eds.). (2012). *Green Oslo. Visions, planning and discourse*. Farnham: Ashgate.

Lund, S., Banta, G. T., & Bunting, S. W. (2014). Applying stakeholder Delphi techniques for planning sustainable use of aquatic resources: Experiences from upland China, India and Vietnam. *Sustainability of Water Quality and Ecology, 3–4*, 14–24.

Luo, S., Wang, H., & Cai, F. (2013). An integrated risk assessment of coastal erosion based on fuzzy set theory along Fujian coast, Southeast China. *Ocean and Coastal Management, 84*, 68–76.

Mahsin, M., Akhter, Y., & Begum, M. (2012). Modeling rainfall in Dhaka division of Bangladesh using time series analysis. *Mathematical Modelling and Application, 1*(5), 67–73.

Matsumoto, J., & Shoji, H. (2003). Seasonal and inter-annual variation of tropical cyclones approaching Vietnam. In *Environmental change and evolution of natural environment in the Red River delta* (pp. 7–60). Japan: University of Tokyo Press. Also available at www.thoitietnguyhiem.net/BaoCao/BaoCaoBao/aspx

Misbahuzzaman, K. (2016). Traditional farming in the mountainous region of Bangladesh and its modifications. *Mountain Science, 13*(8), 1489–1502.

Moges, A., & Holden, N. M. (2007). Farmers' perceptions of soil erosion and soil fertility loss in Southern Ethiopia. *Land Degradation & Development, 18*, 543–554.

Momani, P. E., & Naill, M. (2009). Time series analysis model for rainfall data in Jordan: Case study for using time series analysis. *American Journal of Environmental Sciences, 5*(5), 599–604.

MONRE (Ministry of Natural Resources and Environment), UNDP. (2010). *Build ability to recover: adaptation strategies for livelihoods in coastal area that is most at risk due to the impact of climate change in Central Vietnam.* Report of Research Project Poverty and Environment (in Vietnamese).

MONRE (Vietnamese Ministry of Natural Resources and Environment). (2016). *Climate change and sea level rise scenarios for Vietnam.* NARENCA Publisher, Vietnam, 170 pages.

Mortreux, C., & Barnett, J. (2009). Climate change, migration and adaptation in Funafuti, Tuvalu. *Global Environmental Change, 19*(1), 105–112.

Mukherjee, N., Hugé, J., Sutherland, W. J., McNeill, J., Van Opstal, M., Dahdouh-Guebas, F., & Koedam, N. (2015). The Delphi technique in ecology and biological conservation: Applications and guidelines. *Methods in Ecology and Evolution, 6*, 1097–1109.

Nazareth, M. (1996). *Introduçao à demografia.* Lisboa: Editorial Presença, 185 pages (In Portuguese).

Neirotti, P., De Marco, A., Cagliano, A. C., Mangano, G., & Scorrano, F. (2014). Current trends in smart city initiatives: Some stylised facts. *Cities, 38*, 25–36.

Newton, A., & Weichselgartner, J. (2014). Hotspots of coastal vulnerability: A DPSIR analysis to find societal pathways and responses. *Estuarine, Coastal and Shelf Science, 140*, 123–133.

Nguyen, V. T., Nguyen, T. H., Tran, T., Pham, T. T. H., Nguyen, T. L., & Vu, V. T. (2011). *Climate change and its impacts in Vietnam.* Hanoi, Vietnam: Science and Technology Publishing, 259 pages. (in Vietnamese).

Nguyen, V. T., Dang, T. L., Nguyen, A. T., Tran, D. L., & Hens, L. (2017). Shifting challenges for coastal green cities. *Vietnam Journal of Earth Sciences, 39*(2), 109–129.

Nguyen, A. T., Nguyen, N. T., Luong, T. T., & Hens, L. (2018). Tourism and beach erosion: Valuing the damage of beach erosion for tourism in the Hoi An World Heritage site, Vietnam. *Environment, Development and Sustainability* https://doi.org/10.1007/s10668-018-0126-y

Njoroge, J. M. (2014). An enhanced framework for regional tourism sustainable adaptation to climate change. *Tourism Management Perspectives, 12*, 23–30.

OECD (1993). *OECD environmental indicators: development, measurement and use.* Reference paper, 37 pages.

Omann, I., Stocker, A., & Jäger, J. (2009). Climate change as a threat to biodiversity: An application of the DPSIR approach. *Ecological Economics, 69*(1), 24–31.

Orencio, P. M., & Fujii, M. (2013). A localized disaster-resilience index to assess coastal communities based on an Analytic Hierarchy Process (AHP). *Disaster Risk Reduction, 3*, 62–75.

Paerl, H. W., Valdes, L. M., Joyner, A. R., Peierls, B. L., Piehler, M. F., Riggs, S. R., Christian, R. R., Eby, L. A., Crowder, L. B., Ramus, J. S., Clesceri, E. J., Buzzelli, C. P., & Luettich, R. A. (2006). Ecological response to hurricane events in the Pamlico Sound system, North Carolina, and implications for assessment and management in a regime of increased frequency. *Estuaries and Coasts, 29*(6), 1033–1104.

Paliwoda, S. J. (1983). Predicting the future using Delphi. *Management Decision, 21*(1), 31–38.

Peterson, D. L., & Johnson, D. R. (1995). *Human ecology and climate change: People and resources in the Far North.* Washington: Taylor & Francis, 337 pages.

Pires, I. M., Gibert, M., & Hens, L. (2010). *Studies in human ecology.* Liber amicorum C. Susanne, M. Nazareth, Ph. Lefèvre-Witier (p. 259). Hanoi, Vietnam: Publishing House for Science and Technology.

Prato, T. (2008). Conceptual framework for assessment and management of ecosystem impacts of climate change. *Ecological Complexity, 5*(4), 329–338.
Pritam, C., & Prasenjit, A. (2010). Shoreline change and sea level rise along coast of Bhitarkanika wildlife sanctuary, Orissa: An analytical approach of remote sensing and statistical techniques. *Geomatics and Geosciences, 1*(3), 976–4380.
Quang, D. V., Schreinemachers, P., & Berger, T. (2014). Ex-ante assessment of soil conservation methods in the uplands of Vietnam: An agent-based modeling approach. *Agricultural Systems, 123*, 108–119.
Reed, M. S., Dougill, A. J., & Taylor, M. J. (2007). Integrating local and scientific knowledge for adaptationto land degradation: Kalahari rangeland management options. *Land Degradation & Development, 18*, 249–268.
Regos, A., Ninyerola, M., Moré, G., & Pons, X. (2015). Linking land cover dynamics with driving forces in mountain landscape of the Northwestern Iberian Peninsula. *Applied Earth Observation and Geoinformation, 38*, 1–14.
Rerkasem, K., Yimyam, N., & Rerkasem, B. (2009). Land use transformation in the mountainous mainland Southeast Asia region and the role of indigenous knowledge and skills in forest management. *Forest Ecology and Management, 31*, 25–36.
Reuveny, R. (2007). Climate change-induced migration and violent conflict. *Political Geography, 26*(6), 656–673.
Rowe, G., & Wright, G. (1999). The Delphi technique as a forecasting tool: Issues and analysis. *Forecast, 15*(4), 353–375.
Saaty, T. L. (1980). *The analytic hierarchy process*. New York: McGraw-Hill.
Saaty, T. L. (1990). The analytic hierarchy process in conflict management. *Conflict Management, 1*(1), 47–68.
Saaty, T. L. (1994). How to make a decision: The analytic hierarchy process. *Interfaces, 24*(6), 19–43.
Saaty, T. L. (2001). Fundamentals of the analytic hierarchy process. In D. L. Schmoldt, J. Kangas, G. A. Mendoza, & M. Pesonen (Eds.), *The analytic hierarchy process in natural resource and environmental decision making, Managing forest ecosystems* (Vol. vol 3). Dordrecht: Springer.
Saint-Macary, C., Keil, A., Zeller, M., Heidhues, F., & Dung, P. T. M. (2010). Land titling policy and soil conservation in the northern uplands of Vietnam. *Land Use Policy, 27*(2), 617–627.
Schad, I., Schmitter, P., Saint-Macary, C., Neef, A., Lamers, M., Nguyen, L., Hilger, T., & Hoffmann, V. (2012). Why do people not learn from flood disasters? Evidence from Vietnam's northwestern mountains. *Natural Hazards, 62*(2), 221–241.
Schmidt, R. C. (1997). Managing Delphi surveys using nonparametric statistical techniques. *Decision Sciences, 28*(3), 763–774.
Schowengerdt, R. A. (1997). *Remote sensing: Models and methods for image processing.* Academic Press, 522 pages.
SGG (Suoi Giang Government). (2016). *Annual report for socio-economic development in 2015–2016 of Suoi Giang commune. Official reports.* Yen Bai province: Vietnam, 25 pages (in Vietnamese).
Shamsnia, S. A., Shahidi, N., Liaghat, A., Sarraf, A., & Vahdat, S. F. (2011). *Modeling of weather parameters using stochastic methods* (ARIMA Model) (Case Study: Abadeh Region, Iran). Proceedings of the International Conference on Environment and Industrial Innovation, IPCBEE 12, IACSIT Press, Singapore, pp. 282–285.
Shreve, C. M., & Kelman, I. (2014). Does mitigation save? Reviewing cost-benefit analyses of disaster risk reduction. *Disaster Risk Reduction, 10*, 213–235.
Sikor, T., & Truong, D. M. (2002). Agricultural policy and land use changes in a Black Thai commune of Northern Vietnam, 1952–1997. *Mountain Research and Development, 22*(3), 248–255.
Söderqvist, T., Brinkhoff, P., Norberg, T., Rosén, L., Back, P. E., & Norrman, J. (2015). Cost-benefit analysis as a part of sustainability assessment of remediation alternatives for contaminated land. *Environmental Management, 157*(1), 267–278.

Steiner, F. (2014). *Human ecology: Overview. Reference module in earth systems and environmental sciences, from encyclopedia of ecology, 2008* (pp. 1898–1906). Elsevier Publishing.

STG (Son Thinh Government). (2016). *Annual report for socio-economic development in 2015–2016 of Son Thinh commune. Official reports*. Yen Bai province: Vietnam, 20 pages (in Vietnamese).

Stoffel, M., Wyżga, B., & Marston, R. A. (2016). Floods in mountain environments: A synthesis. *Geomorphology, 272*, 1–9.

Styger, E., Rakotondramasy, H. M., Pfeffer, M. J., Fernandes, E. C. M., & Bates, D. M. (2007). Influence of slash-and- burn farming practices on fallow succession and land degradation in the rainforest region of Madagascar. *Agriculture, Ecosystems & Environment, 119*, 257–269.

Subedi, M., Hocking, T. J., Fullen, M. A., Mccrea, A. R., Milne, E., Wu, B.-Z., & Mitchell, D. J. (2009). Use of farmers' indicators to evaluate the sustainability of cropping systems on sloping land in Yunnan Province, China. *Pedosphere, 19*(3), 344–355.

Subedi, M., Matthews, R. B., Pogson, M., Abegaz, A., Balana, B. B., Oyesiku-Blakemore, J., & Smith, J. (2014). Can biogas digesters help to reduce deforestation in Africa? *Biomass and Bioenergy, 70*, 87–98.

Tastle, W. J., & Wierman, M. J. (2005). *Consensus and dissension: a new measure of agreement*. Fuzzy Information Processing Society 2008. NAFIPS 2008. Annual Meeting of the North American, pp. 1–4 https://doi.org/10.1109/NAFIPS.2005.1548566

TBLG (Thuong Bang La Government). (2016). *Annual report for socio-economic development in 2015–2016 of Thuong Bang La commune. Official reports*. Yen Bai province: Vietnam, 14 pages (in Vietnamese).

TPG (Tran Phu Government). (2016). *Annual report for socio-economic development in 2015–2016 of Tran Phu town. Official reports*. Vietnam: Yen Bai province, 23 pages (in Vietnamese).

Tran, D. L., Hens, L., Cao, T. T. T., & Do, T. T. H. (2014). *Environmental management of seaports in Vietnam* (p. 316). Hanoi: Publishing House for Science and Technology.

Tran, H. T., Tran, P., Hawley, K., Khan, F., & Moenc, M. (2015). Quantitative cost-benefit analysis for typhoon resilient housing in Danang city, Vietnam. *Urban Climate, 12*, 85–103.

UNDP. (2007). *Fighting climate change: Human solidarity in a divided world*. Human Development Report 2007/2008. United Nations, New York, 31 pages.

UN-Habitat. (2009). *Planning sustainable cities: global report on human settlements 2009 by United Nations Human Settlements Programme*. Earthscan/UN-Habitat, London, 2009. No. of pages: xxx + 306. ISBN 978-1-84407-899-8.

VCG (Van Chan Government). (2000). *Report on natural hazard prevention 1998–1999*. Yen Bai, 15 pages (in Vietnameses).

VCG (Van Chan Government). (2002). *Report on natural hazard prevention 2001*. Yen Bai, 5 pages (in Vietnameses).

VCG (Van Chan Government). (2005). *Report on natural hazard prevention 2004*. Yen Bai, 4 pages (in Vietnameses).

VCG (Van Chan Government). (2006). Report on natural hazard prevention 2005. Yen Bai, 18 pages (in Vietnameses).

VCG (Van Chan Government). (2007). *Report on natural hazard prevention 2006*. Yen Bai, 3 pages (in Vietnameses).

VCG (Van Chan Government). (2008). *Report on natural hazard prevention 2007*. Yen Bai, 5 pages (in Vietnameses).

VCG (Van Chan Government). (2009). *Report on natural hazard prevention 2008*. Yen Bai, 6 pages (in Vietnameses).

VCG (Van Chan Government). (2010). *Report on natural hazard prevention 2009*. Yen Bai, 10 pages (in Vietnameses).

VCG (Van Chan Government). (2011). *Report on natural hazard prevention 2010*. Yen Bai, 10 pages (in Vietnameses).

VCG (Van Chan Government). (2012). *Report on natural hazard prevention 2011*. Yen Bai, 12 pages (in Vietnameses).

VCG (Van Chan Government). (2013). *Report on natural hazard prevention 2012*. Yen Bai, 8 pages (in Vietnameses).
VCG (Van Chan Government). (2014). *Report on natural hazard prevention 2013*. Yen Bai, 5 pages (in Vietnameses).
VCG (Van Chan Government). (2015). *Report on natural hazard prevention 2014*. Yen Bai, 6 pages (in Vietnameses).
VCG (Van Chan Government). (2016a). *Report on natural hazard prevention 2015*. Yen Bai, 14 pages (in Vietnameses).
VCG (Van Chan Government). (2016b). *Van Chan Statistic year book 2015*. Yen Bai, 350 pages (in Vietnamese).
Victoria, K., & Turner, S. (2016). Yielding to high yields? Critiquing food security definitions and policy implications for ethnic minority livelihoods in upland Vietnam. *Geoforum, 71*, 33–43.
VNG (Government of Vietnam). (2007). Decision No. 1076/QD-TTg approving the master plan on construction of Vung Ang economic zone, Ha Tinh province, up to 2025. Hanoi Retrieved from http://extwprlegs1.fao.org/docs/pdf/vie76870.pdf
Wang, J., Gao, W., Xu, S., & Yu, L. (2012). Evaluation of the combined risk of sea level rise, land subsidence, and storm surges on the coastal areas of Shanghai, China. *Climatic Change, 115*(3–4), 537–558.
WB (World Bank). (2010). *Natural hazards, unnatural disasters: The economics of effective protection. International Bank for Reconstruction and Development*. Washington, DC: World Bank Publications, 276 pages.
Wilenius, M., & Tirkkonen, J. (1997). Climate in the making: Using Delphi for Finnish climate policy. *Futures, 29*(9), 845–862.
Wilks, A. R. (2013). Preliminary cost benefit analysis of storm surge hazard prevention in the Tuamotu Islands, French Polynesia. SPCSOPAC Technical Report. Natural Resources Economics Section, Ocean and Islands Programme. ISBN 978-982-000643-0.
Yoon, S. T., Lee, Y. H., Hong, S. H., Kim, M. H., Kang, K. K., Na, Y. E., & Oh, Y. J. (2013). Vulnerability assessment of cultivation facility by abnormal weather of climate change. *Korean Journal of Agricultural and Forest Meteorology, 15*(4), 264–272.
Zolingen, S. J., & Klaassen, C. A. (2003). Selection processes in a Delphi study about key qualifications in senior secondary vocational education. *Technological Forecasting and Social Change, 70*, 317–340.
Zurayk, R., Awar, F., Hamadeh, S., Talhouk, S., Sayegh, C., Chehab, A. G., & Shab, K. (2001). Using indigenous knowledge in land use investigations: A participatory study in a semi-arid mountainous region of Lebanon. *Agriculture, Ecosystems & Environment, 86*, 247–262.

Index

A. T. Nguyen, L. Hens, *Human Ecology of Climate Change Hazards in Vietnam*, Springer Climate, https://doi.org/10.1007/978-3-319-94917-8

The manufacturer's authorised representative in the EU is Springer Nature Customer Service Centre GmbH, Europaplatz 3, 69115 Heidelberg, Germany. If you have any concerns regarding our products, please contact ProductSafety@springernature.com

Printed and bound by CPI Group (UK) Ltd, Croydon, CR0 4YY
06/10/2026
02228973-0001